PROJECT AIR FORCE

Valuing Air Force Electric Power Resilience

A Framework for Mission-Level Investment Prioritization

Anu Narayanan, Debra Knopman, Kristin Van Abel,
Benjamin M. Miller, Nicholas Burger, Martha V. Merrill,
Alexander D. Rothenberg, Luke Muggy, Patrick Mills

Prepared for the United States Air Force

Approved for public release; distribution unlimited

For more information on this publication, visit www.rand.org/t/RR2771

Library of Congress Cataloging-in-Publication Data is available for this publication.
ISBN: 978-1-9774-0180-9

Published by the RAND Corporation, Santa Monica, Calif.

© Copyright 2019 RAND Corporation

RAND® is a registered trademark.

Cover: MNF-S forces defend Panama Canal during PANAMAX exercise/TSgt Heather Redman

Support RAND
Make a tax-deductible charitable contribution at
www.rand.org/giving/contribute

www.rand.org

Preface

This report describes how the U.S. Air Force currently sets priorities among its investments in power resilience measures, and how it could improve on current decisionmaking processes to reduce the likelihood that electricity-dependent mission-critical functions are disrupted in the event of a prolonged grid outage. The Air Force's interest is to ensure that it is making the most efficient and productive use of available resources to reduce vulnerability and increase mission resilience in the face of a wide range of power outage scenarios. This report was written as part of research commissioned by Mark Correll, Deputy Assistant Secretary of the Air Force for Environment, Safety and Infrastructure, and conducted within the Resource Management Program of RAND Project AIR FORCE as part of the project entitled Air Force Energy Assurance.

RAND Project AIR FORCE

RAND Project AIR FORCE (PAF), a division of the RAND Corporation, is the U.S. Air Force's federally funded research and development center for studies and analyses. PAF provides the Air Force with independent analyses of policy alternatives affecting the development, employment, combat readiness, and support of current and future air, space, and cyber forces. Research is conducted in four programs: Force Modernization and Employment; Manpower, Personnel, and Training; Resource Management; and Strategy and Doctrine. The research reported here was prepared under contract FA7014-16-D-1000.

Additional information about PAF is available on our website: www.rand.org/paf/

This report documents work originally shared with the U.S. Air Force on February 21, 2018. The draft report, issued on February 28, 2018, was reviewed by formal peer reviewers and U.S. Air Force subject-matter experts.

Contents

Figures

Tables

Summary

Electric power drives command and control, communications, information flows, data storage, heating and air conditioning, and other functions that enable the U.S. Air Force to execute its missions. However, the full value to the Air Force of having assured access to electric power is sometimes not realized until a disruption in power service results in adverse mission effects, sometimes through an unexpected chain of equipment failures across multiple facilities and installations. This report addresses the Air Force's need for a systematic way to value the risk to missions associated with power-related disruptions and accordingly set priorities among investments in power resilience.

Electricity, drinking water, and wastewater systems have historically been treated as installation infrastructure under the responsibility of the base civil engineers (BCEs). Now, the rapidly evolving nature of missions with dependencies on highly reliable power supply and power conditioning to operate sensitive electronic equipment has elevated the importance of energy assurance across the enterprise. These changing circumstances present the Air Force with an opportunity to review and reconsider relevant funding and procurement processes used to allocate resources to new investment and sustainment of electric power capabilities.

Building on previous RAND Project AIR FORCE (PAF) work, this report proposes a framework for assessing the value of power resilience investments for Air Force mission assurance. We first survey best practices and academic literature on valuation approaches and identify desirable attributes of valuation frameworks. Using these attributes as a guide, we then examine how the Air Force currently values power resilience investments and identify areas for improvement. We outline a proposed framework that addresses resource allocation from a mission perspective, focuses on resilience-related investments, considers a broad range of scenarios, and presents cost and performance information that supports risk-informed decisionmaking. Finally, we recommend additional process and policy changes to ensure appropriate valuation of power resilience.

Desirable Attributes of Valuation and Prioritization Processes

We searched for relevant examples of power resilience valuation methods in successful use in organizations outside the Air Force but were unable to identify any promising methods. As an alternative, informed by previous PAF research and a review of the economics literature on valuation methods, we recommend that the Air Force adopt an approach that leverages the essential features of economic valuation. This approach is centered on making informed choices about options and trade-offs by ensuring that investment decisions reflect input from relevant decisionmakers at different levels and the best feasible-to-collect information on the monetary costs and nonmonetary benefits of power resilience investment options. Specifically, we have identified four desirable attributes that should be part of any approach that the Air Force takes in valuing power resilience:

- The focus is on *resilience-related investments*, as distinct from expenditures on maintenance associated with routine operations.
- The analysis takes a *mission perspective*, as distinct from an installation perspective, to properly account for interdependencies across installations that shape mission performance.
- Given the focus on resilience, the analysis has the capacity to evaluate mission performance across a *range of outage scenarios*, not only the ones previously experienced or most expected.
- The analysis presents *key cost and performance information* that forms the basis of a power resilience trade-off framework.

Valuation and Prioritization in Current Air Force Funding Streams

We examine the extent to which the above attributes are seen in current Air Force processes for setting priorities under funding mechanisms commonly used for power-related investments. The Air Force uses several funding streams to procure power resilience options. We looked at three most frequently used sources of funding for power-related investments: the Facilities, Sustainment, Restoration, and Modernization (FSRM) account under the operations and maintenance (O&M) appropriation; military construction (MILCON) under its own appropriations line item; and third-party (private-sector) financing of installation investments facilitated by the Air Force Office of Energy Assurance (OEA). Table S.1 shows the processes used to secure each of the three primary funding streams.

We also looked for evidence of these four attributes in current Air Force processes for setting priorities in the three most relevant funding streams: FSRM, MILCON, and third-party financing. We found that while these processes reflect some of the

Table S.1
Air Force Processes and Funding Streams for Power Resilience Investments

Funding Stream	Air Force Process
FSRM account under the O&M appropriation	Air Force Comprehensive Asset Management Plan (AFCAMP) process managed by the Air Force Installation and Mission Support Center (AFIMSC) and executed by the Air Force Civil Engineering Center (AFCEC)
MILCON appropriation	Planning and programming for MILCON projects process owned by Headquarters, Air Force, Logistics, Engineering, and Protection (HAF/A4), with final prioritization for Energy Resilience and Conservation Investment Program (ERCIP) projects done at the Office of the Secretary of Defense (OSD) level across U.S. Department of Defense (DoD) components
Third-party financing (e.g., power purchase agreements, utility energy savings contracts, and energy performance savings contracts)	Not tied to one specific named process, this is the primary mechanism used by OEA.

NOTE: OSD typically does not edit the normal MILCON project lists submitted by each component but does hold final decisions, explicitly, on projects submitted for the ERCIP, a subset of MILCON funding. (Email communication with DoD subject-matter expert, February 8, 2018).

desired attributes, none exhibits all of the attributes needed for the Air Force to make effective power resilience investment decisions.

Proposed Valuation Framework

The ultimate goal for the Air Force is to compare the value of power resilience investments for one mission against different types of power- and non-power-related investments for other missions. As a step toward that larger goal, this analysis focuses on the immediate problem of how the Air Force can set priorities among power resilience–related investments only. Nested within this problem are two subproblems: (1) prioritizing power-related investments for *one mission carried out across multiple installations* and (2) prioritizing power-related investments that span *multiple missions and installations.* The framework presented in this report addresses the former.

The Air Force currently does not have a systematic way to choose among competing power resilience options for known or anticipated performance gaps for a given mission. We present a seven-step framework that provides a way for the Air Force to identify the preferred power resilience options for a single mission carried out across multiple installations. The proposed framework is designed to support decisions about different power resilience options and not to determine whether to pursue routine maintenance procedures.[1]

We use the mission of the 618th Air Operations Command (AOC), also known as the Tanker Airlift Control Center (TACC),[2] to illustrate data availability and interpretation of performance. The 618th AOC's mission is to plan, task, execute, and assess global mobility operations. Several aspects of TACC's mission rely heavily on assured access to electric power, making TACC a useful example against which to test the energy valuation framework proposed in this report. Our objective in using the TACC example is to provide concrete illustrations of outputs that users of the framework could expect to get out of each step. Our objective is *not* to arrive at "the best" power resilience measures for TACC or to suggest that what works for TACC would work for other Air Force missions. We use a hypothetical scenario to walk through framework steps that are scenario-dependent, and we provide an overview of data and information gaps that we encountered as we worked through the TACC example.

The framework consists of seven analytical steps, some requiring data-gathering and others requiring analysis and subjective assessments. Figure S.1 provides a schematic of the steps in the framework. Step 1 includes a process by which the mission owner defines an appropriate metric for measuring mission performance and then specifies the desired or targeted performance for the mission using this metric.[3] As part of this step, the *status quo* power architecture, procedures, and personnel are also reviewed. In Step 2, the mission owner, with help from BCEs, chooses relevant scenarios that will be used in the analysis to stress the mission and identify vulnerabilities. While there are no "right" test scenarios, framework users should select scenarios that span a wide range of outage conditions and challenge any assumptions made about resources and plans available to cope with power disruptions. We discuss this critical framework step in detail in Chapter Four. The analysis in Steps 3 through 6 is carried out for each of the individual scenarios and entails the estimation of costs and performance under each scenario if no further power resilience options were to be implemented (we call these *status quo costs* and *performance*, respectively); costs and performance for each scenario under a set of power resilience options; and an articulation of preferences based on those expected costs and performances. Finally, Step 7 looks across the outage scenarios to support selection of a broadly applicable power resilience option for the mission. As detailed in the body of this report, the framework incorporates the four desired attributes that we identified as necessary for Air Force valuation

---

[1]   Basic maintenance is critical to ensuring that systems are operating as intended. Without standard maintenance, the likelihood of system dysfunction or failure can significantly increase.

[2]   We use "618th AOC" and "TACC" interchangeably in this report.

[3]   We broadly define the term *mission owner* as Narayanan et al. (2017) do: "Mission owners are commanders of organizations responsible for some kind of operational outcome. A mission could be fighter pilot training; operating and maintaining unmanned aerial systems; or operating a radar installation, command headquarters, or tenant unit for another service or government agency" (p. 16).

Figure S.1
Framework for Identifying Preferred Power Resilience Option for One Mission

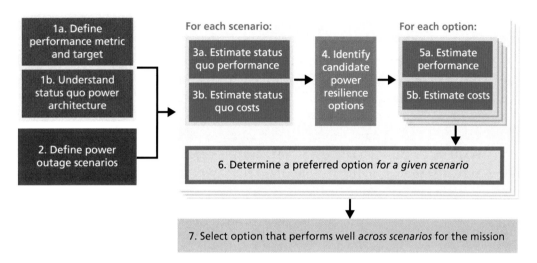

and prioritization of power resilience investments. The framework's individual steps are illustrated to the extent possible by using the 618th AOC's mission and notional data.

The proposed framework can be incorporated into three Air Force funding processes in which power resilience investments are valued and prioritized. The first is to prioritize projects tagged as "energy assurance" in the future AFCAMP process.[4] The second is to prioritize projects within the MILCON ERCIP. The third is to guide the selection process for resilience investments made using third-party financing mechanisms through, for example, OEA. While evaluation of such projects considered by OEA is primarily based on the potential for cost reductions, the presented framework can help to ensure that resilience investments also formally account for mission risk.

Recommendations

We recommend the Air Force take the following steps, in order of their degree of difficulty in implementation:

- **Collect data required to implement the proposed framework.** Cognizant of the burden that any data collection effort imposes on organizations, we nonethe-

---

[4] For the FY 2019–2021 cycle, installations are being asked to tag projects that have an energy assurance component to them—for example, investments that reduce single point(s) of failure vulnerabilities or improve energy security in some way—so the Air Force can begin tracking investments that support energy resilience goals enterprisewide (Air Force Civil Engineer Center Planning and Integration Directorate, undated).

less find that data within easy reach of gathering and compiling could improve the quality of the analysis embedded in the valuation framework. Several of the data gaps we identified (detailed later in this report) could be easily filled by recording information that is spread among many within the enterprise in one place.

- **Increase BCEs' views into mission owners' contingency plans (e.g., continuity of operation plans).** This would enable BCEs to be better positioned to maintain a mission perspective when identifying and budgeting for installation needs.

- **Create a forum for evaluating incremental versus transformational power resilience options for a given mission.** Whereas power resilience options are currently sought in different funding streams, this recommendation would draw attention to expanding the range of potential power resilience options, regardless of which funding stream is pursued.

- **Use standard asset management principles for routine maintenance and repairs.** Routine maintenance is foundational to power resilience and should be prioritized at the highest levels of the Air Force.

- **Consolidate the budgeting and management of electricity and communications systems in investment and asset management.** These systems are increasingly interdependent; consolidation would expand the options for building resilience and efficiency into these systems and would streamline procurement, construction, and maintenance.

- **Consider cost-sharing to encourage mission owners to internalize more of the costs associated with reliable and resilient power for their missions.** Mission owners could communicate to AFCEC a clearer signal of their willingness to share in the costs of increasing resilience and enable the Air Force's power resilience resources to be allocated more effectively across missions.

- **Consider power resilience in acquisition and strategic basing processes.** By addressing power resilience needs earlier in the life cycle of systems, solutions could be more efficient, affordable, integrated, and creative. Considering an installation's vulnerabilities to a diverse set of power outage scenarios as part of the strategic basing process could ensure that locations that offer an appropriate level of power resilience are selected for a given weapon system.

The framework and recommendations presented in this report are aimed at improving the Air Force's ability to make risk-informed decisions about power resilience investments. The goals of the framework are to identify and present decision-relevant information in a clear manner that facilitates decisionmaking. A necessary first step in prioritizing power-related investments for a given mission, and eventually across missions, is to provide decisionmakers with a means of weighing performance improvements afforded by a power resilience option against the costs of implementing it. While many of these investment decisions are cost-sensitive, there may be situations in which the cost of power resilience measures is not an objective (or is small relative to the cost of the asset at risk), implying that national security demands whatever measures are necessary to ensure an uninterruptible power supply under the most challenging and rare outage scenarios.

The framework does not require monetizing performance benefits. It does, however, involve making a necessarily subjective assessment of how much a given improvement in performance is worth. The proposed valuation framework does not take away the need for the decisionmaker's judgment. Rather, it simply structures and presents information that a mission owner or other mission-oriented decisionmaker needs to understand the mission risks posed by power outages and, when appropriate, to trade those risks against the costs involved in mitigating them.

Having a valuation framework alone does not guarantee sound investment decisionmaking. Incorporating asset management principles and best practices, improving communications between key stakeholders and aligning their incentives, and considering power resilience in multiple processes are all essential components of a robust energy valuation strategy.

Acknowledgments

We are grateful for the support of our sponsors Mark Correll, Deputy Assistant Secretary of the Air Force for Environment, Safety, and Infrastructure, and to Doug Tucker for his guidance throughout the course of the project.

We also would like to thank various people across the Air Force, including base civil engineers and mission owners at Vandenberg Air Force Base and Scott Air Force Base (particularly members of the 618th Air Operations Command), for engaging with the RAND Project AIR FORCE team and providing us with information that served as the basis for our analysis.

Within the RAND Corporation, we are grateful to Obaid Younossi for his program leadership and project guidance, and also to our RAND colleagues Paul Davis, Ryan Consaul, and Quentin Hodgson for their critical reviews of the report. In addition, John Conger, former Principal Deputy Under Secretary of Defense, served as an outside reviewer and provided many constructive comments about the practicality of the proposed framework.

Abbreviations

ACES	Automated Civil Engineering System
AFB	Air Force Base
AFCAMP	Air Force Comprehensive Asset Management Plan
AFCEC	Air Force Civil Engineering Center
AFI	Air Force Instruction
AFIMSC	Air Force Installation and Mission Support Center
AoA	analysis of alternatives
AOC	Air Operations Command
APA	additional performance attribute
BCA	benefit-cost analysis
BCAMP	Base Comprehensive Asset Management Plan
BCE	base civil engineer
CBA	capabilities-based assessment
CCMD	combatant command
CDD	capabilities development document
CE	civil engineering
CEA	cost-effectiveness analysis
COF	consequence of failure
CONOPS	concept of operations
COOP	continuity of operations
DAS	Defense Acquisition System
DoD	U.S. Department of Defense
DOTMLPF-P	doctrine, organization, training, materiel, leadership and education, personnel, facilities, and policy
EMD	Engineering and Manufacturing Development
EPSC	energy performance savings contract

ERCIP	Energy Resilience and Conservation Investment Program
FOC	full operational capability
FSRM	facilities sustainment, restoration, and modernization
FY	fiscal year
GDSS	Global Decision Support System
HAF	Headquarters, Air Force
HAF/A4	Headquarters, Air Force, Logistics, Engineering, and Protection
HAF/A4C	Headquarters, Air Force, Logistics, Engineering, and Force Protection, Directorate of Civil Engineers, Facilities Division
HVAC	heating, ventilation, and air conditioning
IC	installation commander
ICD	initial capability document
IPL	Integrated Priority List
JCDIS	Joint Capabilities Integration and Development System
JUON	joint urgent operational need
KPP	key performance parameter
KSA	key system attribute
LBNL	Lawrence Berkeley National Laboratory
LSP	National Aeronautics and Space Administration Launch Services Program
MAIS	Major Automated Information System
MAJCOM	major command
MCAMP	MAJCOM Comprehensive Asset Management Plan
MDA	Milestone Decision Authority
MDAP	Major Defense Acquisition Program
MDD	Materiel Development Decision
MDI	Mission Dependency Index
MILCON	military construction
MS	Mission Profile
MSA	Materiel Solutions Analysis
NASA	National Aeronautics and Space Administration
NFPA	National Fire Protection Association
NPD	National Aeronautics and Space Administration Policy Directive

NPR	National Aeronautics and Space Administration Procedural Requirement
O&M	operations and maintenance
O&S	Operations and Support
OEA	Air Force Office of Energy Assurance
OSD	Office of the Secretary of Defense
PAF	RAND Project AIR FORCE
PoF	probability of failure
PPA	power purchase agreement
PPBE	planning, programming, budget, and execution
PSC	public service commission
PUC	public utility commission
RDT&E	research, development, test, and evaluation
SAF/IE	Assistant Secretary of the Air Force for Installations, Environment, and Energy
SMS	Sustainment Management System
TACC	Tanker Airlift Control Center
UESC	utility energy savings contract
UFC	unified facility code
UPS	uninterruptible power supply

CHAPTER ONE

Introduction

All U.S. Air Force missions need assured access to energy, especially electric power, at installations. Electric power drives command and control, communications, information flows, data storage, heating and air conditioning, and other functions that enable the Air Force to execute its missions. However, the full value to the Air Force of having assured access to electric power is sometimes not realized until a disruption in power service adversely affects missions, sometimes through an unexpected chain of equipment failures across multiple facilities and installations. This report addresses the Air Force's need for a systematic way to value risk to missions associated with power-related disruptions and accordingly to set priorities among investments in power resilience.

Power Resilience as an Exercise in Risk Management

U.S. military installations depend on the commercial electric grid. The grid is predominantly under private control but regulated through state public service commissions (PSCs) and the Federal Energy Regulatory Commission. While the grid has been made much more reliable in the past decade or so through aggressive efforts of the consortium of grid operators under the umbrella of the North American Electric Reliability Corporation, it remains vulnerable to catastrophic outages (North American Electric Reliability Corporation, 2018). Threats to the grid come in many forms. Extreme weather and natural disasters might be increasing in frequency and severity. For instance, in 2012, Hurricane Sandy caused widespread power outages and disrupted natural gas and petroleum distribution systems for weeks in parts of the Northeast (U.S. Department of Energy, 2013). Additionally, aging infrastructure is increasingly a concern, as some of the grid's components are more than 100 years old:

> Most transmission and distribution lines were constructed in the 1950s and 1960s with a 50-year life expectancy, and were not originally engineered to meet today's demand, nor severe weather events." (Gerrity and Lantero, 2014)

In 2003, a combination of human and equipment failures caused 50 million people in the Northeast to lose power for up to two days (Minkel, 2008). Modernization of the grid to incorporate supervisory control and data acquisition (SCADA) control systems creates potential cybersecurity vulnerabilities. In 2015, Ukraine power companies suffered a targeted cyberattack that caused only a brief grid power outage, but it took months for the companies to recover from damage to controls systems (Zetter, 2016).[1] Finally, seemingly unrelated events, such as physical attacks, might be of concern simply because they are out of the control of grid operators. For instance, in 2016, power was cut to Incirlik Air Force Base (AFB) for almost a week during a Turkish coup attempt (Schogol and Palwyk, 2016).

The remote locations of many military bases, aging on-base power infrastructure, and issues with unreliable electric power grids on bases outside of the continental United States all make U.S. Department of Defense (DoD) installations especially vulnerable to power outages (Marquesee, Schultz, and Robyn, 2017). While the external threat cannot always be stopped or mitigated, the Air Force does have some control over its vulnerability to these threats and associated consequences. Diesel-powered backup generators, uninterruptible power supply (UPS) systems, and continuity of operations (COOP) plans typically provide resilience against grid outages. Fuel for generators and off-base maintenance personnel are often key components of on-base resilience capabilities. In addition to being able to physically access the base, engineers and maintenance personnel need a significant understanding of the status quo architecture so that they can effectively identify and fix problems as they arise.

The Air Force cannot control threats to electric power; it also cannot pursue every possible power resilience option. Therefore, the challenge is to have a transparent, empirically grounded way to understand the value of different power resilience options. In the absence of such understanding (which we call a "power resilience valuation framework" hereafter), Air Force decisionmakers do not know whether they are investing the right amount of money in the appropriate mix of power resilience options, given their risk tolerance.

Research Questions

The Air Force has commissioned two prior RAND Project AIR FORCE (PAF) projects on aspects of the power resilience problem. The first was a fiscal year (FY) 2016 project that focused on the development of an installation energy assurance assessment framework. As part of a second project, PAF provided analytic support to the Air Force Mission Assurance Tiger Team as it developed a 24-month action plan for assessing

---

[1]   The short outage duration can be largely attributed to the ability of workers to operate breakers and other elements of the Ukrainian power grid in manual mode. Power grid controls systems in the United States typically lack such manual backup functionality, making power restoration a more difficult and longer task.

vulnerabilities and adaptive capacities in the face of catastrophic power outages. The research presented in this report is focused on four questions:

- How should the Air Force evaluate power resilience investments, drawing on the economics literature and standard practice among a select number of public and private sector examples?
- How well do current Air Force processes evaluate power resilience investments?
- What is a recommended valuation framework, and how can it be applied?
- What additional issues should the Air Force consider in improving its valuation of power resilience?

Definitions

We define *resilience* as the ability of a system to withstand and recover from a disruption. Resilience is often discussed in the context of high-consequence, low-probability events, such as natural disasters or determined attacks. We define *robustness* as the ability to withstand and recover from multiple types of disruptions or scenarios. A system may be subjected to any number of different types of disruption scenarios. No system can be robust across all imaginable scenarios (Narayanan et al., 2017). Similarly, we define *robust resilience options* as those that are effective in the face of a diverse set of disruption scenarios.

Finally, we define *electric power[2] resilience investments*, which are the focus of this work, as expenditures on measures to improve the resilience of a mission to disruptive power outages. *Power resilience options* could take any number of forms so long as they help to mitigate power-related risks to mission performance in one way or another. They could include the addition of redundant power sources or distribution infrastructure (e.g., adding a backup generator, adding redundant commercial electrical lines); diversification of the generation mix (e.g., adding renewables or other onsite generation); modernization of the electrical infrastructure (e.g., installing microgrid technologies); or establishment of facilities at other locations at which the mission could be carried out (e.g., adding an alternative location for contingency operations or continuity of operations, the latter of which is known as a COOP site).

For purposes of this analysis, we also distinguish between routine maintenance to support normal operations and such investments as upgrading or modernizing heating, ventilation, and air conditioning (HVAC) systems from the more specialized and targeted investments intended to provide resilience to power outages. However, it is important to note that spending on routine maintenance to maintain all equipment in good working order, whether power-related or not, is foundational to building resil-

---

[2]   We use "electric power" and "power" interchangeably in the rest of this report.

ience and should be considered a necessary condition in advance of more specialized investments.

Approach and Organization of This Report

Chapter Two considers traditional economic approaches to valuation. We sought to learn from non-DoD organizations about their energy assurance processes and how they plan for and cope with power disruptions.[3] Our intent was to gather insights that could either be directly applied to Air Force processes or provide additional context as to how other organizations attempt to solve issues pertaining to energy assurance. We also reviewed past PAF work and the scholarly literature primarily to ground the broader discussion of valuation and identify directions the Air Force could go in the future.

In Chapter Three, we examine the processes for valuation and priority setting currently used by the Air Force to allocate resources to power resilience investments. For the Air Force, valuing electric power—and by extension, valuing the operational attributes of power resilience—means understanding the appropriateness and priority of investments. The Air Force routinely exercises priority-setting in the context of its funding stream for major military construction (MILCON), facilities sustainment, restoration, and modernization (FSRM), as well as more specialized funding mechanisms to support energy assurance. As a starting point for our analysis, we sought to understand how power resilience investments were valued in these processes. To understand conditions on the ground and constraints faced by mission owners and base civil engineers (BCEs) in identifying, funding, and implementing power resilience options, we visited two installations—Vandenberg AFB in California and Scott AFB in Illinois—each with critical power needs but each with distinct missions and tolerances for prolonged outages.[4] The purpose of our visits was to enable us to understand key decisionmaking processes, sustainment protocols, and the two missions' sense of their own vulnerabilities to outages.

Our interest is in understanding the strengths and weaknesses of these valuation processes under conditions of uncertainty about the timing, nature, and severity of the outage, and designing an analytical approach that would better serve the needs of the Air Force. We recognized that trying to quantify risk to mission is nearly impossible

---

[3] We limited our survey of best practices to non–DoD organizations (as opposed to all organizations) upon the sponsor's request.

[4] We broadly define the term *mission owner* as Narayanan et al. (2017) do: "Mission owners are commanders of organizations responsible for some kind of operational outcome. A mission could be fighter pilot training; operating and maintaining unmanned aerial systems; or operating a radar installation, command headquarters, or tenant unit for another service or government agency" (p. 16).

to do in absolute terms given the many uncertainties associated with outage scenarios. Instead, we reframed the valuation challenge around collecting and presenting relevant cost and performance information in such a way that a decisionmaker with knowledge of mission needs and priorities can identify the preferred power resilience options for reducing mission disruptions resulting from electric power outages.

In Chapter Four, we present our proposed valuation framework and illustrate each step with an example drawn from a hypothetical case based in part on data obtained from the Tanker Airlift Control Center (TACC) located at Scott AFB.

In Chapter Five, we step back from the details of the framework to identify larger concerns with the current Air Force approach to power resilience that affect its ability to implement the proposed valuation framework. We then offer recommendations to address these issues. Appendixes provide additional background information on risk perception, approaches to valuation taken by non-DoD organizations, and upstream processes for setting and modifying power resilience needs.

CHAPTER TWO

Approaches to Valuation and Desirable Attributes of a Valuation Framework

In this chapter, we review various sources that informed the development of a valuation framework for the Air Force. First, we describe our efforts to review methods and practices of other organizations. We then review previous PAF research focused on Air Force energy assurance to provide insights on how value assessments might be conducted specifically in the Air Force context. We also consider how the concept of "value" is treated in the economics literature[1] and applied in various economic sectors to ground the broader discussion of valuation and inform future directions for the Air Force.[2] We conclude the chapter by identifying four attributes that should be present in any Air Force assessments of power resilience investments. We use these attributes to evaluate current Air Force prioritization processes (Chapter Three) and to develop an improved valuation framework (Chapter Four).

Review of Non-DoD Practices

Before embarking on an effort to develop a practical valuation framework for the Air Force, we wanted to first look at methods and practices of other organizations who cope with their vulnerabilities to power outages. To guide our selection of relevant organizations, we identified a small set of organizational attributes that capture essential features of the Air Force that are most relevant to its dependence on electricity and its needs for high reliability and power quality. The attributes we chose, in no particular order, were:

---

[1]   This chapter is not meant to provide an exhaustive review of alternative approaches to valuing investment options; rather, we describe the standard approaches commonly used to compare investment options and common methods for valuing benefits of different options to explain our reasoning for selecting the four attributes we identify as important to Air Force assessments of power resilience investments.

[2]   We also provide a brief summary of lessons learned through our review of non-DoD practices related to the valuation of energy assurance in Appendix B.

- **Distributed locations:** Air Force missions are carried out at many locations throughout the United States and abroad and thus require a high degree of networking of communications, missions, and supporting functions, such as logistics and personnel management.
- **Nested levels of decisionmaking:** The Air Force has a hierarchical organizational structure, with investment decisions being made at all levels.
- **Diverse set of missions:** Air Force missions vary widely, from traditional flying missions to intelligence-gathering missions to humanitarian aid and disaster relief support. We therefore sought organizations that have distinct, varied, and interrelated missions that have different power needs.
- **Formal procurement processes:** DoD (and, thus, the Air Force) uses highly structured and formalized budgeting and procurement processes.
- **Effect on lives lost:** Disruption to Air Force missions may carry serious consequences for national security and the protection of human lives.
- **Recent power disruption:** This attribute does not relate to organizational makeup but, for purposes of this research effort, we wanted to focus on organizations that may have recently changed or amended policies in response to a power disruption.

Given the unique mission and organization of the Air Force, we knew that no organization would embody all of these attributes. Nonetheless, it was helpful to use these attributes as a loose guide in our selection process. We interviewed representatives from six organizations that share a subset of these attributes with the Air Force—the National Aeronautics and Space Administration (NASA), two hospital systems, a university system, an electric utility, and a regional transmission organization.[3] We also considered the practices of the catastrophic insurance industry. A detailed description of each of these case studies can be found in Appendix B.

Although no single organization or corresponding sector offered a direct analog to the Air Force, we gained an appreciation for how power resilience is interpreted by other organizations and found the exercise useful in sharpening our understanding of the defining characteristics of the problem for the Air Force. The organizations we examined tended not to focus on reliability and resilience until they experienced major power outages. Their planning efforts are driven largely by past experience and tend to take a limited view of power outage scenarios that have not been experienced. Given these shortcomings, we concluded that the Air Force could not simply borrow from other organizations' methods and experience. Consequently, this exercise in the end did not contribute to the development of the framework for analysis described in Chapter Four.

---

[3]   See Appendix B for a map of attributes to organizations.

Review of Recent PAF Studies for the Air Force on Energy Assurance

Our assessment of Air Force energy assurance valuation was informed by insights from past PAF analyses. For example, a baseline understanding of the Air Force's vulnerabilities to catastrophic or long-duration outages is fundamental to the process of choosing desirable investments. An earlier PAF analysis considered the analytical problem of assessing vulnerability of individual Air Force installations to catastrophic power outages (Narayanan et al., 2017). The focus of that research was on systematic measurement of baseline conditions of vulnerability across a range of outage scenarios that typically are not considered in Air Force planning. The framework for baseline vulnerability analysis across scenarios could then be applied to estimate changes in vulnerability as a consequence of applying power resilience options, both materiel and operational in nature. The report made various recommendations and concluded that a framework that

> incorporates clearly and simply defined metrics; clear guidance on roles, responsibilities, and necessary communication channels; and a systematic way to think through vulnerabilities (i.e. assessing the effect of adverse scenario conditions on system architecture elements) can go a long way toward making risk-informed decisions when it comes to energy assurance. (Narayanan et al., 2017, p. 52)

The report also noted that decisionmakers should "look across missions at a given base and across bases that support a particular mission before investing in energy assurance upgrades" (Narayanan et al., 2017, p. 53).

A second phase of PAF work provided analytical support to an Air Force Tiger Team tasked with developing a 24-month action plan to systematically assess the nature and extent of vulnerabilities at mission-critical nodes using a progression of discussion-based and operational exercises at selected locations, drawing on the framework to provide a consistent analytical approach. The challenge was to take a mission-centric view (rather than an installation-centric one) of risk as a means of defining and scoping the problem of reducing the vulnerabilities of Air Force missions to catastrophic power outages. The report outlined a plan with four lines of effort: (1) identify critical nodes for analysis, (2) map the system architectures of these nodes under normal conditions, (3) conduct exercises to test vulnerability and adaptive capacities of these nodes, and (4) institutionalize the lessons of the action plan for future assessments. Execution of the plan is under way and will provide the Air Force with a better picture of its enterprisewide needs to enhance energy assurance under a broader range of scenarios than typically used. The valuation framework described in Chapter Four of this report calls for a similar range of exercises to understand the performance of current and potential power resilience measures under various outage scenarios.

This analysis represents a third phase of PAF work, in which we shift our focus from assessing vulnerability to valuing power resilience, thus enabling comparison of the benefits and costs associated with the Air Force's many decisions about whether, what, and how to spend its resources on enhancing assured access to power *across installations for a single mission across a range of extended, disruptive outage scenarios.*[4] Such a valuation method is fundamental to the priority-setting processes that the Air Force employs in its budgeting for energy-related projects.

In examining DoD and Air Force funding mechanisms for energy-related projects, we have discovered that routine maintenance investments and power resilience investments often compete for the same funds. While investments in routine maintenance are foundational to power resilience and have implications for the sustainment of infrastructure assets, the same criteria cannot be used for prioritizing both maintenance work and resilience investments. Investments in routine maintenance should focus on safeguarding against known and common failure modes to ensure that reliability standards are met. Probabilistic risk assessment approaches can inform investment decisionmaking when it comes to routine maintenance. The same cannot be said about decisions regarding power resilience investments that need to account for low probability, high-consequence events. The likelihood with which such events might occur is highly uncertain, diminishing the usefulness of traditional risk assessment methods in this context.

Common Approaches to Comparing Investment Choices

Valuation fundamentally involves a comparison of benefits and costs. Benefits and costs can take on various forms and be compared in different ways. In this section, we present a high-level overview of common approaches to making investment choices and consider how the concept of "value" is addressed in the economics literature. The goal of our literature review was to identify approaches that seem most suitable for the Air Force power resilience investment decisionmaking context.

Economics literature suggests two main approaches to investment or planning decisions that the Air Force faces. The first is a benefit-cost analysis (BCA) framework, which compares the cost of a choice with the benefits that choice provides, with both terms denominated in dollars. In most cases, the costs are relatively easy to calculate, especially when the cost is primarily the money needed to purchase and maintain a system. However, monetizing the benefits can be a much more difficult task. Furthermore, it does not always make sense to monetize benefits. For example, in the case of power resilience, the benefits of an investment may include enhanced mission perfor-

---

[4]   This analytical problem is more challenging than setting priorities among power-related investments *for one mission at one installation and for a single scenario.* The even more challenging problem to tackle is a credible analytical approach to allocating resources *across installations and missions.* We do not solve this problem in this report. Instead, we set our sights on the intermediate goal of demonstrating a practical approach to setting priorities among power-related investments *for one mission carried out across multiple installations.*

mance in the form of reduced downtime or delays, reduced need for other investments, or even lives saved. To denominate these benefits in dollar terms would be challenging and may lead to investment choices that do not reflect what the decisionmaker truly values. For this reason, a BCA framework is not appropriate for making Air Force power resilience investment decisions.

An alternative approach to BCA is a cost-effectiveness analysis (CEA), in which a decisionmaker compares a set of options to determine which provides the desired level of effectiveness at the lowest cost. A cost-effectiveness decision criterion is sometimes described as seeking the "biggest bang for the buck." The advantage of CEA is that it can be used to assess benefits that are not denominated in dollars. However, CEA cannot be used to determine whether any one option provides benefits greater than its cost, as long as the benefits are in units other than dollars. Furthermore, simple CEAs assume that two options that are equally cost-effective hold equal value, irrespective of either option's overall capacity for performance improvement. In practice, other metrics, such as reduction of mission delays and total cost may be needed to discriminate between options. For example, suppose Option 1 reduces mission delays by six hours each time a particular power outage scenario occurs and costs $120,000 annually, while Option 2 reduces mission delays by four hours each time a particular power outage scenario occurs and costs $80,000 annually. For a scenario frequency of twice annually, the cost per hour of reduced mission delay (the cost-effectiveness metric) for both options is $20,000. Under these conditions, mission owners with a relatively high tolerance for power outages would likely favor spending less overall. But mission owners with a low tolerance for any duration of outage might favor Option 1 over Option 2 and be willing to pay the extra $40,000 to reduce the outage duration by an additional two hours.

The valuation framework presented in Chapter Four of this report favors neither a direct comparison of costs and monetized benefits (as is done in a BCA), nor an approach that selects the option with the lowest ratio of costs to effectiveness as the winner (as is done in a CEA). Rather, the framework aims to build a trade-space of cost and performance information for a candidate set of power resilience options under a range of power outage scenarios. This information can help Air Force decisionmakers make trade-offs among costs, mission performance, and associated mission risks for a single mission.

Both BCA and CEA approaches use a single measure of effectiveness against which to compare options. In reality, decisionmakers often desire to meet multiple objectives at once and are subsequently faced with the problem of choosing among options that are scored against different measures of effectiveness. There are several approaches available to tackle these types of problems, including portfolio analysis techniques, scorecards, and other forms of multiobjective decisionmaking.[5] However, the framework presented in this work focuses on helping Air Force mission owners and BCEs identify preferred power resilience options for *one mission* by presenting information regarding the costs and performance of a set of candidate power resilience options for a range of power outage scenarios. As such, the framework assumes that mission owners will be able to find an appropriate measure of effectiveness (or performance metric) against which to measure success. If the framework were expanded to enable cross-mission decisionmaking, it would need to handle multiple measures of effectiveness because the same performance metric cannot be used to measure the success of every Air Force mission.

Methods for Valuing the Benefits of Investment Choices

A separate challenge from choosing an appropriate valuation approach across both benefits and costs is settling on a method by which to value the *benefits* associated with investment options. There is a robust economics literature on valuation that provides a range of methods to estimate the value of difficult-to-measure concepts, such as power resilience. The two main types of methods used to estimate broad-based benefits, effectiveness, or performance of investments are known as *revealed-* and *stated-preference* methods.

Revealed-Preference Methods

Revealed-preference methods rely on existing markets and the goods traded within them to help determine the value of a given benefit (Boyle, 2003). For example, to understand how valuable power resilience is, we could look at the market for backup generators. One of the implicit "goods" that consumers purchase when they buy a backup generator is resilience to grid outages. The cost of owning and operating generators could be used as a proxy for the value of resilience. Revealed-preference methods rely on some type of observed market data or behavior to calculate or "back out" the value of a good or an attribute of a good.

In practice, interpreting market valuations requires an assumption that buyers participating in the market are unbiased and act rationally. For the case of power resilience, a further assumption would need to be made that buyers have an accurate sense of failure probabilities (which they probably do not). These assumptions are unlikely to be valid in a national security environment. Appendix A describes several examples of psychological and behavioral biases that complicate the interpretation of findings from a revealed-preference approach.

---

[5] For example, Davis, Shaver, and Beck (2008) and Davis and Dreyer (2008) have developed the Portfolio Analysis Tool, which helps analysts and decisionmakers assess portfolios of investment options using a hierarchy of policy scorecards that can be drilled into interactively to understand how the options were scored.

The Air Force's Integrated Priority List (IPL) might seem like a good source of information on Air Force leadership's revealed preferences pertaining to power-related investments. The IPL contains data on Air Force purchases of energy-related equipment and services for all infrastructure-related FSRM projects. The IPL is compiled and maintained by the Air Force Civil Engineering Center (AFCEC) and is updated annually as part of the budget process. An understanding of how the Air Force has historically chosen to prioritize power-related investments, in principle, could serve as an important baseline against which to compare future investments. However, there is a danger of leaning too heavily on past investment patterns to inform future decisionmaking. As noted earlier, properly interpreting IPL and other historical data on investment choices requires an assumption that decisionmakers are unbiased and have an accurate sense of the power-related risks to missions. Additionally, the threats and hazards facing the Air Force, the Air Force's mission set, and the option set available to mitigate risks are in a near-constant state of flux. Any future decisions about power resilience investments ought to take into account this evolving landscape.

Stated-Preference Methods

In contrast to revealed-preference approaches, stated-preference methods are useful when a good is not traded in a market, even indirectly. In these cases, we ask individuals or organizations questions that allow us to assess the value of the benefit in question (Brown, 2003). High-quality stated-preference methods rarely ask people how much they are willing to pay directly for a good. Instead, the questions allow the analyst to infer willingness to pay from hypothetical choices. A classic example is air quality, which is not typically traded in a market but has value. To measure how consumers value clean air, we can ask a series of questions focused on trade-offs between air quality and other goods, including money, and calculate the implicit value of having clearer air. This is an example of the most common stated-preference method, *contingent valuation,* which is sometimes used as a general term for stated-preference approaches. The last two steps of the framework presented in Chapter Four use a similar tactic to elicit decisionmaker preferences.

Stated-preference methods involve formal surveys, interviews, tabletop exercises, or some other form of engagement with relevant decisionmakers. This engagement can be time-consuming and requires sufficient buy-in from those who are expected to participate. Additionally, stated-preference methods rely on hypothetical choices, which are less reliable indicators of preference than actual choices. However, for some valuation activities, such as the one in this work, stated-preference methods are the only available option for gaining a deeper understanding of the benefits side of the equation.

Desirable Attributes of Valuation and Priority-Setting Processes

Informed by previous PAF research and a review of the economics literature on valuation methods, we recommend that the Air Force adopt an approach that leverages the essential features of economic valuation, specifically making informed choices, grounded in analysis, about options and trade-offs. The Air Force can operationalize this approach by ensuring that investment decisions reflect input from relevant decisionmakers at different levels, and that they draw on feasible-to-collect information about the monetary costs and nonmonetary benefits of power resilience investment options. Specifically, we have identified four desirable attributes that should be part of any approach the Air Force takes to valuing power resilience:

- The focus is on *resilience-related investments*, as distinct from expenditures on maintenance associated with routine operations.
- The analysis takes a *mission perspective*, as distinct from an installation perspective, to properly account for interdependencies across installations that shape mission performance.
- Given the focus on resilience, the analysis has the capacity to evaluate mission performance across a *range of outage scenarios*, not only the ones previously experienced or most expected.
- The analysis *presents key cost and performance information* that forms the basis of a power resilience trade-off framework.

In Chapter Three, we examine the extent to which these attributes are seen in current Air Force processes for setting priorities under funding mechanisms commonly used for power-related investments.

Current Air Force Approaches to Power Resilience Investments

The previous chapter identified the attributes that one would wish to see in an assessment of Air Force power resilience investments. This chapter examines whether current processes—especially those associated with the most common funding streams—embody those attributes and where improvements could be made.

For context, it is useful to note the distinct roles and responsibilities of civil engineering (CE) staff on installations and mission owners who are tenants of the installations. The CE staff are responsible for daily base operations and for ensuring that the infrastructure-related mission needs are met. They perform these duties by planning and programming basic support functions, developing military construction and other infrastructure projects, and submitting these plans for consideration through the appropriate funding streams. Installation support staff have various funding streams available to them from the normal annual planning, programming, budgeting, and execution (PPBE) process to centralized sustainment funds and construction funds to alternative financing mechanisms with nongovernment funds.

Mission owners also have a number of funding streams available to them for operating (e.g., fuel for aircraft), acquiring (e.g., equipment to support mission functions), and operations and maintenance (O&M; typically used as support for contractors). In some cases, mission owners may wish to spend their funds on buying power resilience measures, although mission owners typically look to installation support staff to take care of these needs.[1] One exception to this is the purchase, operation, and maintenance of UPS systems, which are considered equipment, which the mission owner is responsible for and uses contractor support to maintain.[2] Mission owners might occasionally use O&M funds, procurement funds, or research, development, test, and evaluation (RDT&E) funds to purchase generators, but this is not the normal path for procuring and installing generators.

---

[1] Each funding stream has certain restrictions on how the money can be used. Large infrastructure projects do not fit neatly into the funds available to mission owners.

[2] In very few cases, CE staff might perform basic maintenance on UPSs if there is a memorandum of agreement between the mission owner and CE. Typically, these functions are performed by contractors.

Because CE staff are responsible for base operations and maintaining installation infrastructure, we focus on the three sources of funding available to them and used most frequently[3] for power-related investments: the FSRM account under O&M appropriation;[4] MILCON under its own appropriations line item; and third-party (private-sector) financing of installation investments facilitated by the Air Force Office of Energy Assurance (OEA) and other organizations within the Air Force. Table 3.1 shows which processes are used to secure each of the three primary funding streams.

In the following sections, we describe in more detail the processes that underlie each funding stream and how well each incorporates the desired attributes of valuation described at the end of Chapter Two.

Facilities Sustainment, Restoration, and Modernization

Overview

The FSRM account is part of the O&M appropriation. This is commonly referred to as "3400 money" in the Air Force, which references the congressional appropriation account. This is the most flexible fund available for making investments in energy

Table 3.1
Air Force Processes and Funding Streams for Energy Resilience Investments

Funding Stream	Air Force Process
FSRM account under O&M appropriation	Air Force Comprehensive Asset Management Plan (AFCAMP) process managed by the Air Force Installation and Mission Support Center (AFIMSC) and executed by AFCEC
MILCON appropriation	Planning and Programming for MILCON projects process owned by Headquarters, Air Force, Logistics, Engineering, and Protection (HAF/A4), with final prioritization for Energy Resilience and Conservation Investment Program (ERCIP) projects done at the Office of the Secretary of Defense (OSD) level across DoD components
Third-party financing (e.g., power purchase agreements [PPAs], utility energy savings contracts [UESCs], energy performance savings contracts [EPSCs])	Not tied to one specific named process, this is the primary mechanism used by OEA

NOTE: OSD typically does not edit the normal MILCON project lists submitted by each component but does hold final decisions, explicitly, on projects submitted for ERCIP, a subset of MILCON funding (email communication with DoD subject-matter expert, February 8, 2018).

---

[3] Other funding streams (such as the procurement appropriation, RDT&E, and mission O&M) may be used but to a lesser extent.

[4] CE staff use the standard DoD programming process to plan and program basic base operations and some routine maintenance and repair activities that fall under the installation's local execution authority. Larger projects are submitted for consideration of centralized funding under the FSRM account.

infrastructure. To secure these funds, installation support organizations can submit projects through the AFCAMP process for consideration of centralized funds managed by AFIMSC and executed by AFCEC.[5]

Before FY 2016, projects seeking centralized FSRM funds were prioritized at the major command (MAJCOM) level. Starting in FY 2016, the Air Force realigned sustainment funding from the MAJCOMs to AFIMSC. With this realignment, all projects eligible[6] for these sustainment funds are now incorporated into one IPL[7] at the Air Force enterprise level by AFCEC. To score and rank projects, AFCEC uses a prioritization model[8] that assigns points to the project in three categories: the probability of failure (PoF), which is a function of the asset's condition index (a value that takes into consideration an asset's age and its useful life); the consequence of failure (CoF), which is a function of the Mission Dependency Index (MDI) and MAJCOM priority points; and estimated cost savings (applicable to energy projects). [9] In theory, the scoring model should prioritize the most critical projects to performing a mission, increasing the likelihood that these projects will fall above the funding cutline.

A simplified version of the AFCAMP process is presented in Figure 3.1. Starting at the top left of the figure, AFCEC uses data from various databases and tools, such as the Automated Civil Engineering System (ACES)[10] and Sustainment Management System (SMS)[11] to prepare a prospectus for each installation. The prospectus contains information on the condition and risks of the installation's infrastructure (Air Force Civil Engineer Center Planning and Integration Directorate, 2015).

The installation uses the prospectus and other installation-specific plans to validate the data, and compile and score projects using the AFCAMP prioritization

---

[5] For a more detailed discussion of Air Force infrastructure funding processes, see Mills et al. (2017).

[6] There is no minimum dollar threshold for a project to be eligible for placement on the IPL. Eligibility is determined by work classification type. See Air Force Instruction (AFI) 32-1032 (2015) for a description of work classification types.

[7] At the time of this writing, the projects being prioritized for funding are programmed for FY 2019 to FY 2021, and the IPL has moved to a three-year rolling list (previously covering two years), in an attempt to start planning further into the future.

[8] The AFCAMP prioritization model is a risk-based model, in which "risk" is *probability multiplied by consequence*; however, from FY 2015 to FY 2017, the model added, instead of multiplied, these elements together. This has the consequence of skewing priorities to projects with high scores in either the probability or consequence term. For example, a project with a high CoF but a low to medium PoF might score higher in an additive model than a project with medium to high score for both factors. This is not the case when you multiply the factors together.

[9] Energy projects wishing to compete on energy savings are assigned additional points based on a savings-to-investment-ratio–tiered structure.

[10] The Air Force is currently transitioning installations to a new real property system called TRIRIGA.

[11] The SMS is a suite of software applications developed by the Engineer Research and Development Center's Construction Engineering Research Laboratory to help facility managers make decision on how best to maintain the built environment.

Figure 3.1
Simplified Representation of the AFCAMP Process

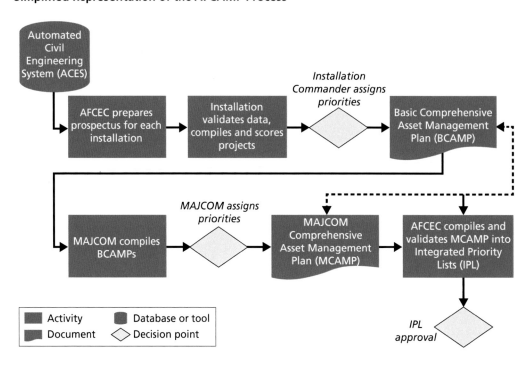

model. All projects and their scores go into the Base Comprehensive Asset Management Plan (BCAMP). Before finalizing the BCAMP, the installation commander (IC) has the opportunity to signal to the MAJCOM which projects should be prioritized. Each installation then sends its completed BCAMP to the MAJCOM, where they are compiled into one plan, the MAJCOM Comprehensive Asset Management Plan (MCAMP). Before finalizing the MCAMP, the MAJCOM commander has the ability to assign MAJCOM priority points to projects.

Finally, all the MCAMPs are sent to AFCEC, where they are compiled into the enterprisewide IPL. Dotted lines in Figure 3.1 between AFCEC compiling the enterprisewide IPL and the boxes that represent development of the BCAMP and MCAMP indicate possible iteration that happens before the final list is complete. The final IPL is approved by HAF/A4.

Not every project included in the IPL gets funded. The threshold is based on the funding level allocated to the centralized FSRM account in the President's Budget each year. Projects that fall below the threshold are not removed from the IPL; instead, these projects stay on the list and are rolled into next year's cycle.[12] Some of these projects might receive "fall out" money at the end of the FY from projects that were funded

---

[12] Some installations may also seek other sources of funds, such as mission-appropriated O&M funds or third-party financing.

in the current FY but failed to execute. However, in interviews with subject-matter experts, we learned it can be difficult to use "fall-out" money on projects that might benefit mission owners and the installation the most. For example, we heard at one installation that current Air Force contracting processes and local authority thresholds prevented the CE organization from using "fall out" money on replacing power poles because there was not enough time to bid the project and do the work before the end of the FY. Instead, they used the money to replace carpets and complete other minor projects because costs associated with these projects were within their local execution authority.

We also learned in our conversations with base personnel that, in building base project lists, there does not appear to be much opportunity to compare maintenance of current electrical infrastructure as is (allowing for some modernization) with bigger transformational changes that could significantly improve resilience through, for example, complete replacement of aging infrastructure with modern systems. That is, strategic goals and short-term operational planning might at times be misaligned. For example, CE personnel at Vandenberg AFB must maintain the South Vandenberg Power Plant as long as it exists, but there is some question as to whether it is still needed at all.

Assessment

The FSRM prioritization model embodies some of the attributes we have identified as desirable for Air Force power resilience valuation, but it falls short in others, as we discuss next.

Focus on Resilience

As with many processes and decision tools, the FSRM prioritization model is continuously being updated. For the FYs 2019–2021 IPL cycle, installations are being asked to tag projects that have an energy assurance component to them, which begins to incorporate a focus on energy resilience (Air Force Civil Engineer Center Planning and Integration Directorate, 2017). However, these projects will still be scored with the current prioritization model, which does not incorporate the desirable attributes for energy resilience valuation as outlined in this assessment section.

Mission Perspective

There are two components of the FSRM prioritization model that incorporate some dimensions of the mission perspective: the MDI and MAJCOM priority points. The intent of MDI is to capture the importance of different assets to a particular mission. An asset's MDI is based on a category code—a five or six-digit code used by DoD that represents a specific type of facility or infrastructure—and is formally factored in the CoF element in the prioritization model. However, the MDI has not been implemented in a manner in the Air Force that accurately captures mission criticality for all assets. For example, a runway is less important to a space mission than it is to a flying mission; however, based on current guidance, these assets will be assigned the same MDI. When asset classifications do not accurately reflect their importance to the mission, the final prioritized list of projects may not reflect the best investments for the Air Force at the mission or enterprise level. To begin to address some of these inconsistencies, AFCEC has implemented an MDI adjudication process whereby installations can submit a request to have a particular asset's MDI reevaluated (Air Force Civil Engineer Center Planning and Integration Directorate, 2015). This is a new effort, and, therefore, we are unable to assess progress or evaluate the effort in this study.

MAJCOM priority points allow MAJCOM commanders to identify and assign points to priority projects. This is a mechanism that MAJCOM commanders can use to move projects higher up on the IPL that, for whatever reason, did not score well but might be critical for supporting a mission.[13] These points are formally factored into the CoF element in the prioritization model. Presumably, MAJCOM commanders will assign priority points to projects that are most critical to carrying out the mission; however, if the MDI accurately reflected the mission criticality of an asset, then the role of MAJCOM priority points (as intended by the model) would become obsolete. We do not suggest that there should be no role for MAJCOM priority-setting in the model; rather, we suggest that the Air Force assess the intent of each element and ensure that its use in the model reflects that intent. For example, in budget-constrained environments, it is probable that not all high-priority projects can be started in the same year. MAJCOM priority points could be used to determine which projects must be performed in the current year versus future years.

Range of Scenarios

There is no Air Force requirement to use a range of outage scenarios in investments decisions. The current prioritization model does formally treat PoF, a logical place to account for the likelihood of various hazards that might affect mission performance; however, this element uses only the condition index of the asset as the influencing factor. Other factors (e.g., natural hazards and manmade threats) can also affect the probability of failure, so while the condition index is an appropriate measure to use for determining investments in routine maintenance and repair projects, it should not be the only criterion used to score power resilience projects. There is, however, a new effort under way at the DoD level to develop an energy resiliency planning unified facility code (UFC), which would require more than one scenario be used to evaluate installation energy resilience.[14] While considering two scenarios is taking a step in the right direction, care should be taken to select scenarios that sufficiently stress the systems being evaluated and challenge commonly made assumptions about plans,

---

[13] It may be the case that a mission has a unique requirement that cannot be captured by the current scoring model.

[14] Email communication with Air Force subject-matter expert, January 25, 2018.

processes, and resources available to cope with power outages. It is unclear to which processes this new UFC will apply, but presumably any process used to evaluate and secure funding for resilience investments will be subject to the requirements and provisions in the UFC, including the AFCAMP process.

Inform Cost-Performance Trade-Offs

Formal mission performance metrics do not play a significant role in the FSRM prioritization model. Projects with estimated costs above certain dollar thresholds require an economic analysis as part of the justification;[15] however, these analyses follow traditional cost-benefit approaches or cost-effectiveness approaches, which do not lend themselves well to difficult-to-measure attributes (such as resilience), as discussed in Chapter Two.

Military Construction

Overview

New construction projects with estimated costs that exceed $2 million are normally funded through the MILCON appropriation (Public Law 115-91, 2017). A portion of the appropriation (the ERCIP) is set aside for energy projects.[16] This program previously only considered projects with energy and water savings, but the program has been expanded to include projects with a focus on energy resilience and energy security in future FYs (Office of the Assistant Secretary of Defense, Energy, Installations, and Energy, 2017).

MILCON projects are identified and developed at the installation level by the IC and CE staff and submitted to MAJCOMs to be prioritized (AFI 32-1021, 2016; AFI 32-1023, 2017). Ultimately, the Secretariat has the final say on budget submissions to Congress, and MILCON projects are detailed in the budget request. DoD used to have the flexibility to determine which projects to approve when the funds were appropriated, but Congress now requires a list of projects that are specifically appropriated. OSD typically does not edit the MILCON project lists submitted by each component but does hold final decisions, explicitly, on projects submitted for MILCON ERCIP.[17]

---

[15] In accordance with AFI 65-501 (2018), projects seeking to commit resources to a new project or program with total investment higher than $2 million or annual recurring costs of more than $500,000 in a FY for at least four years require an economic analysis. Repair projects for which the estimated costs are equal to or greater than 75 percent of the replacement value also require an economic analysis.

[16] This portion is expected to be around $150 million through FY 2023 (Office of the Assistant Secretary of Defense, Energy, Installations, and Energy, 2017).

[17] Email communication with DoD subject-matter expert on February 8, 2018.

Assessment

Focus on Resilience

The expansion of MILCON ERCIP to include energy resilience and energy security projects is a recognition that these projects should be valued and scored separately from conservation projects. The FY 2019–2020 ERCIP memorandum directed to DoD components states that "submissions should focus on proposed projects that would not necessarily be candidates for third party financing or Operations and Maintenance (O&M) funds" (Office of the Assistant Secretary of Defense, Energy, Installations, and Energy, 2017). However, project prioritization is subjective and performed at the DoD level. Furthermore, the pot of money is relatively small, and thus few projects will eventually be funded through this stream.

Mission Perspective

The mission perspective is partially incorporated into the MILCON process, with the MAJCOMs prioritizing projects (presumably mission-critical projects). However, because there is no formal requirement to assess mission needs at an installation, projects are typically developed with consideration only for the requirements of the portion of the mission performed at that installation.

MILCON ERCIP, specifically, gives some consideration of each proposed project's contribution to mission assurance. The top criteria on which projects are evaluated is the "degree to which projects improve an installation's energy resilience and contribute to mission assurance," followed by service priority, project savings, incorporation into installation plans, technology considerations, and the degree to which projects also contribute to conservation (Office of the Assistant Secretary of Defense, Energy, Installations, and Energy, 2017).

Range of Scenarios

As stated above in the FSRM assessment section, there is no Air Force requirement to use a range of scenarios in investments decisions. However, the MILCON process will presumably also be subject to the requirements and provisions of the new energy resilience planning UFC, when released.

Inform Cost-Performance Trade-Offs

MILCON projects are prioritized at the MAJCOM level, but we are not aware of any formal treatment of mission performance in that prioritization process. MILCON ERCIP guidance does state that energy resilience and energy security projects must be cost-effective, but it is unclear how a project's contribution to mission performance will be valued.

Similar to the FSRM funding stream, projects above a certain dollar threshold within the MILCON program require justification based on economic analysis,[18] but again, these analyses are based on traditional cost-benefit approaches.

Third-Party Financing

Overview
Third-party financing is another way that installation support staff can try to fund energy projects. These are such contracts as PPAs, EPSCs, and UESCs. PPAs are contracts between two parties in which one party produces power and the other party agrees to purchase that power at a certain rate for the term of the contract, typically for 10 to 20 years. PPAs are commonly used to finance the installation and operation of solar photovoltaic arrays, microturbines, fuels cells, and other distributed energy generation assets. EPSCs and UESCs are partnerships between either a private energy service company (in the case of EPSCs) or a commercial utility (in the case of UESCs) in which work is performed to upgrade or modernize building electrical systems with no upfront capital required by the user. The costs of the project are paid for by the energy savings generated by the improvements. This is the primary funding stream that OEA uses to fund and execute power resilience options at select installations (OEA, 2018).

Assessment
Focus on Resilience
This attribute is not applicable to third-party financing. These financing mechanisms were created to help spur investment in cost-effective energy conservation measures but do not formally treat resilience in the calculation, and, therefore, only energy resilience projects that promise energy savings are typically funded using these mechanisms.

Mission Perspective
There is no formal treatment of the mission perspective with third-party financing, and the emphasis on cost-reduction may not necessarily support the mission perspective.

Range of Scenarios
As stated earlier, there is no Air Force requirement to use a range of scenarios in investments decisions. However, presumably projects seeking funding through third-party financing will also be subject to the requirements and provisions of the new energy resilience planning UFC, when released.

---

[18] New construction with estimated costs higher than $2 million requires an economic analysis (AFI 65-501, 2017).

Inform Cost-Performance Trade-Offs

Similar to the other funding streams, there is no formal treatment of mission performance with third-party financing. The emphasis of these mechanisms is primarily on cost reduction.

Summary of Findings

These funding streams and processes incorporate some but not all of the desirable attributes for valuation and priority-setting processes, as summarized in Table 3.2. First, the focus on resilience is being incorporated into both the FSRM and MILCON funding streams. For the FY 2019–2021 IPL cycle, installations have been asked to tag energy assurance projects, and the MILCON ERCIP was expanded in 2018 to include energy resilience and energy security projects. Second, the mission perspective is incorporated into the FSRM funding stream with MDI and MAJCOM priority points and, to some extent, into the MILCON funding stream; third-party financing mechanisms do not formally treat the mission perspective. Third, no funding stream formally requires evaluation of a range of outage scenarios in investment decisions. Finally, information necessary to trade costs for performance in a risk-informed way is not incorporated into any of the three funding streams. These funding streams use traditional approaches to valuation and priority setting, such as CBA.

Drawing from this discussion, in Chapter Four, we present our proposed valuation framework for power resilience investment decisions, which incorporates all four of the desirable attributes of valuation and priority-setting processes.

Table 3.2
Summary of Assessment of Funding Streams and Attributes

	FRSM	MILCON	Third Party
Mission perspective	• MDI used to prioritize • MAJCOM input reflects mission perspective • *MDI does not accurately capture mission criticality*	• MAJCOM input reflects mission perspective • *Project justifications made at installation level with no formal requirement for mission perspective*	• *Emphasis on cost reduction does not necessarily support mission assurance*
Range of scenarios	• Effort to incorporate scenarios into new energy resilience planning UFC • *No Air Force requirement to evaluate range of scenarios in investment decisions*		
Inform cost-performance trade-offs	• *Mission performance metrics do not play a significant role*	• *No formal treatment of mission performance in MAJCOM prioritization*	• *No formal treatment of mission performance*
Focus on resilience	• Installations to tag energy assurance projects for FY 2019–2021 AFCAMP • *Condition indexes should not drive valuation of resilience projects*	• MILCON ERCIP attempts to prioritize projects that improve resilience • *Prioritization is subjective and done at DoD enterprise level*	N/A

NOTE: Normal text represents a positive assessment. Italicized text represents a critique.

CHAPTER FOUR
Proposed Power Resilience Valuation Framework

The ultimate goal for the Air Force is to be able to trade power resilience for one mission against some other type of nonpower investment for another mission. As a step toward that larger goal, this analysis focuses on the immediate problem of how the Air Force can set priorities among power resilience–related investments only. Nested within this smaller problem are two subproblems: (1) prioritizing power resilience–related investments *for one mission carried out across multiple installations*, and (2) prioritizing power resilience–related investments that span *multiple missions and installations*. The framework presented in this chapter addresses the former, as the Air Force currently does not have a systematic way to choose among competing power resilience measures for known or anticipated performance gaps for a given mission. The chapter concludes with a brief and high-level discussion of potential approaches to addressing the latter.

We anticipate the proposed framework being useful in at least three ways. The first is to prioritize projects tagged as "energy assurance" in the future AFCAMP process.[1] The second is to prioritize projects within the MILCON ERCIP. The third is to guide the selection process for resilience investments made using third-party financing mechanisms through, for example, OEA. While evaluation of such projects is primarily based on the potential for cost reduction, the presented framework can help to ensure that resilience investments considered in this stream formally account for a primary goal of reducing mission risk.

This chapter presents a seven-step framework that provides a way for the Air Force to identify the preferred power resilience options for a single mission carried out across multiple installations. The proposed framework is designed to support decisions about different power resilience options and not to determine whether to pursue routine maintenance procedures. Although resource constraints always mean careful thought is required for determining what investments are appropriate, basic mainte-

---

[1] For the FY 2019–2021 cycle, installations are being asked to tag projects that have an energy assurance component to them—for example investments that reduce single point(s) of failure vulnerabilities or improve energy security in some way—so the Air Force can begin tracking investments that support energy resilience goals enterprise wide (Air Force Civil Engineer Center Planning and Integration Directorate, 2017).

nance is critical to ensuring that systems are operating as intended. Without standard maintenance, the likelihood of system dysfunction or failure can significantly increase.

We use the mission of the 618th Air Operations Command (AOC), or TACC, to illustrate data availability and interpretation of performance.[2] TACC, located at Scott AFB, is considered a global weapon system consisting of nine directorates that "plan, schedule, and direct a fleet of more than 1,300 mobility aircraft in support of combat delivery and strategic airlift, air refueling and aeromedical evacuation operations around the world" (618th Air Operations Center, undated). Grid power to Scott AFB comes from two utility providers, Ameren and Direct Energy. Our objective in using the TACC example is *not* to arrive at "an answer" for TACC. Rather, where possible, we use the TACC mission as a way to provide concrete examples of outputs that users of the framework should expect to get out of each step. We use a hypothetical scenario to walk through all framework steps that are scenario-dependent. We conclude the framework walkthrough by providing an overview of data and information gaps we encountered as we worked through the TACC example.

Overview of Framework Steps

The framework consists of seven analytical steps, some requiring data-gathering and others requiring analysis and subjective assessments. Figure 4.1 provides a schematic of the steps of the framework. Step 1 includes a process by which the mission owner defines an appropriate metric for measuring mission performance and, using this metric, specifies the desired or targeted performance for the mission. Specifying this

Figure 4.1
Framework for Identifying Preferred Power Resilience Option for One Mission

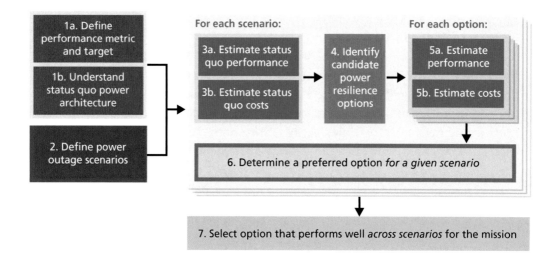

---

[2] We use "618th AOC" and "TACC" interchangeably in this report.

targeted performance is not a straightforward task. For example, suppose that a mission is carried out across multiple installations or involves activities carried out in different organizations that work together to accomplish the mission. In this case, this step requires the mission owner to have a significant understanding of each supporting installation or organization's contribution to the overall mission's performance. Furthermore, there might be several targeted performance thresholds (not just one) for a mission, depending on the operating conditions. What constitutes a performance target for a "blue skies" day might not accurately represent the desired performance on a day when significantly off-normal demands are placed on the mission. Also, as part of this step, the *status quo* power architecture, procedures, and personnel are reviewed.

In Step 2, the mission owner, with help from BCEs, chooses a set of relevant scenarios that will be used in the analysis to stress the mission and identify vulnerabilities. The analysis in Steps 3 through 6 is carried out for each of the individual scenarios and entails the estimation of costs and performance under each scenario if no further power resilience options were to be implemented (we call these *status quo costs* and *performance*, respectively); costs and performance for each scenario under a set of power resilience options; and an articulation of preferences based on those expected costs and performances. Finally, Step 7 looks across the chosen scenarios to support selection of a broadly applicable power resilience option or set of options for the mission.

Proposed Activity Owners for Each Step

We propose that different stakeholders own different steps of the framework, with the Assistant Secretary of the Air Force for Installations, Environment, and Energy (SAF/IE) overseeing the coordination of its implementation. Table 4.1 suggests activity owners for each step.

Step-by-Step Description

Step 1a: Define Performance Metric and Targeted Performance
General Description

As an enabling asset, the value of energy is driven by the value of the mission itself. The first step in understanding energy's role in mission performance is having a way to measure mission performance using a tracked or trackable performance metric. The appropriate metric will vary from one mission to another; for example, one might choose the number of sorties generated per day or the percentage of plans completed daily. Because the value of energy is related to avoided degradations in mission performance, it may be more useful to select a metric that measures mission degradation, such as average delays or the number of cancellations or missed performance targets.

Table 4.1
Suggested Activity Owner for Each Step in the Framework

Framework Step	Suggested Activity Owner
1a. Define performance metric and targeted performance	Mission owners*
1b. Understand status quo power architecture	BCEs
2. Define power outage scenarios	BCEs, mission owners* (with help from external subject-matter experts)
3a. Estimate performance under status quo power architecture	BCEs, mission owners*
3b. Estimate costs under status quo power architecture	BCEs, mission owners*
4. Identify candidate power resilience options	BCEs with support from some central Air Force function (e.g., AFCEC mission energy assurance cell)
5a. Estimate performance for each power resilience option	BCEs, mission owners*
5b. Estimate costs for each power resilience option	BCEs with support from some central Air Force function
6. Determine a preferred option for a given scenario	Mission owners*
7. Select option that performs well across scenarios for the mission	Mission owners*

* If the mission is carried out at multiple sites, seek input from mission owners at all sites.

A mission owner might use historical performance as a proxy for targeted performance during a power outage scenario and use the framework to test the extent to which performance during the scenario deviates from this baseline level. Alternatively, if past performance is not considered an appropriate measure of desired performance, then the mission owner might specify a performance target (using the performance metric chosen in this step) against which to measure mission performance during each scenario of interest. However, what is considered acceptable performance on a regular day might not pass on a day when there are significant demands placed on the mission. In specifying this performance target, the mission owner might decide to set the bar high by assuming that the scenario occurs during an already stressing time for the mission. For the TACC case, this might mean that there are two hurricanes requiring humanitarian relief while there is also an ongoing presidential move. For other missions, a "wartime" operational tempo might determine the performance target.

Another option for the mission owner would be to consider a few different targeted performance thresholds. If multiple performance targets are chosen, then the first six steps of the framework would need to be completed for each.

Whichever way the mission owner chooses to proceed, the two outputs of Step 1a remain the same: a relevant mission performance metric, and one or more targeted performance thresholds measured using this performance metric. The aim is for the Air Force to understand the relationship between (1) the target and achievable performance levels with various power resilience options and (2) the target and performance-cost trade-offs.

We use "targeted performance" rather than "requirement" to describe the output of this step for two reasons. First, the "requirement" might not always be clear. In this case, rather than guessing at the requirement, the mission owner can select a performance target to walk through the steps of the framework to gain an understanding of what it costs to meet this target. If the power resilience option selected in Step 7 comes at a higher cost than expected, this might motivate a discussion of whether the selected performance target is too high, or whether there might be a lower performance threshold that would better reflect the mission owner's risk tolerance. Alternatively, a clear "requirement" might exist, but the mission's success might not be measured relative to this requirement. Rather, going above and beyond this requirement might be the norm for the mission. In this case, the framework can help mission owners identify the point of diminishing returns beyond which marginal improvements to performance are not worth the cost.

TACC Example

To illustrate Step 1a, we began by selecting an appropriate metric for measuring the performance of the TACC's mission. Figure 4.2 provides a simplified description of the workflow associated with TACC's mission. Users submit requests for mobility support. TACC confirms requirements associated with these requests and determines whether resources such as aircraft and crews are available to meet each requirement. Once feasibility is confirmed, aircraft are assigned to the mission and logistical details are planned. As long as no unexpected issues arise, the assigned aircraft then execute the mission as planned.

TACC provided PAF with data from the Global Decision Support System (GDSS) that document total mission execution delays on a daily basis.[3] Delay data that are

Figure 4.2
TACC Mission Life Cycle

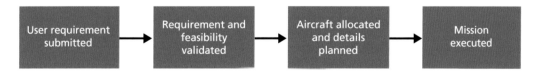

---

[3] We define *delay* as the difference between the initially planned execution time and the actual execution time. This metric is likely to underestimate total delays because it does not account for delays in the pre-execution phase of the mission that might result from a power outage or other types of disruptions.

broken up by mission phase (requirement submission, validation, aircraft allocation and planning, or execution) or reported at the individual mission level (rather than summed up across missions) might add value, but the provided data serve as a useful starting point. Figure 4.3 shows actual total daily hours of mission execution delays across all executed missions for TACC in a three-month period.

Our intent in walking through the steps of the framework is not to come up with an "answer" for TACC but rather to illustrate the process and identify potential challenges to implementation. For this reason, we assume that historical (or baseline) performance is a reasonable stand-in for targeted performance. This assumption may not be appropriate for other missions for which historical conditions fail to adequately capture the stresses under which mission assurance should be evaluated.

A key takeaway is that TACC has an established way to measure and track performance. This might not be the case for every mission in the Air Force enterprise. For those missions that currently cannot do the same, the first order of business will be to define and establish a way to track appropriate performance metrics.

Step 1b: Understand Status Quo Power Architecture
General Description
Before making decisions about desired changes to the status quo system in later steps, it is important to understand how the system currently operates. For energy assurance, the status quo system includes not only the relevant power architecture, but also relevant contingency and continuity of operations (e.g., COOP) plans and personnel

Figure 4.3
Total Hours of Delay Across Missions Executed on a Given Day

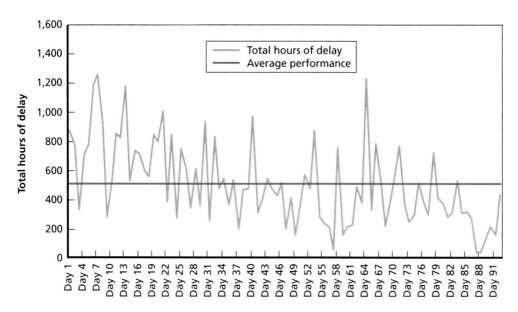

required for both normal and contingency operations (e.g., contracted support personnel, off-base maintenance personnel). There should be an understanding of which backup systems are currently in place, such as generators and uninterruptible power supply systems, as well as the details of contractual agreements and contingency plans that are activated when a power outage occurs. This understanding of the status quo system is critical to support the framework's later steps, such as identifying feasible power resilience options or estimating costs.

Users of this framework should be sure to gather information about status quo power architecture costs during normal grid operations, such as the cost of purchasing grid power, as well as the costs of owning, maintaining, and operating backup power generators, UPS systems, and other power resilience measures. This understanding of status quo costs will serve as a reference point against which to compare the costs associated with power resilience options as part of later steps in the framework. Other status quo costs to track include those associated with personnel, maintenance contracts, and fuel delivery.

The proposed framework focuses on a specific mission and requires a comparison of status quo electricity costs for this mission against the costs of implementing one or more additional power resiliency measures. However, a potential challenge to assessing status quo electricity costs for a particular mission is that these costs might only be tracked at the installation level. Individual mission owners generally do not incur the costs of purchasing grid power, or own and operate backup power generators. These expenses are typically paid from installation CE funds and tracked at the installation level. If a mission-level breakdown of these costs is not available, BCEs and other energy planners on an installation will need to generate some type of estimate of mission-specific costs. As installations move toward tracking electricity usage and generator runtimes at the facility level, it will become easier to isolate status quo power system costs at the mission level.

For this exercise, it is particularly important to look at the costs that might be subject to change under different power resilience options, as the results of this framework will hinge on differences in performance and cost between different power resilience options. Most costs identified in this step will be relatively constant annual costs, such as the purchase of energy from the grid and the maintenance of on-base infrastructure. Later steps will focus on additional costs incurred when an outage occurs.

TACC Example
We were able to obtain data on the cost of grid-supplied electric power for Scott AFB as a whole from November of 2016 through September of 2017, although we do not know what fraction of these costs reflect TACC's power usage. Having at least a full year of data is important for estimating normal costs of grid power, as it is well known that electricity expenditures vary seasonally, although this variation is larger for residential

consumers than commercial or industrial consumers (Energy Information Administration, 2013).

We were also able to obtain data on some costs associated with the number of primary and backup generators at Scott AFB, their approximate replacement costs, the number devoted to TACC, and the hours of run time provided by the associated tank of fuel. Furthermore, we learned that in the event of an outage scenario, the BCE has the ability to allocate a small number of additional mobile generators across the base as needed. We were unable to access data on the costs associated with maintenance, fuel, or other associated personnel costs, but this does not mean that these data do not exist. Backup generators are typically maintained by the BCE and through maintenance service agreements, which often require an individualized quote.[4] Scott AFB, like many Air Force bases, also has contracts for the delivery of fuel, the details of which are not publicly available.

Irrespective of who on an installation incurs these costs (e.g., BCEs or mission owners), from the Air Force enterprise perspective all of these costs should be included. For comparability and use in future steps of this framework, these costs should also be amortized over the life of each component to reflect the average cost per year.

Step 2: Define Power Outage Scenarios
General Description

The next step is to identify several relevant stressing scenarios to examine the resilience of the mission and supporting systems. Scenarios should span a range of durations, from hours to months, and spatial scales, from installation-level to regional. To serve as examples, in previous work (Narayanan et al., 2017), the five scenarios summarized in Table 4.2 were developed to cover a wide range of possible outage conditions defined in part by their temporal and spatial scales, as shown in Figure 4.4.[5] The frequency of these or virtually any other scenarios is highly uncertain and, in many cases, unknowable, which is the primary reason why a formal probabilistic approach to risk assessment is neither appropriate nor feasible. Repeated mechanical failures that occur with observable (and high) frequency may be an exception.

Table 4.3 describes each scenario dimension listed in the first column of Table 4.2. The most challenging scenarios push beyond a few days to a week and the standard operating conditions for diesel generators. Practitioners can adjust both the effects associated with a scenario and the narrative for the scenario to consider situations that are relevant for their location, or that are anticipated to degrade mission performance

---

[4]   Maintenance agreements for smaller backup generators can be generalized. For a few examples, see Parra Electric (undated) and NNG Standby Automatic Generators (undated) and NGG Automatic Standby Generators (undated).

[5]   In addition to these scenarios, as users work through the steps of the framework, they should remain open to discovering other, more stressing scenarios that might degrade mission performance more than is acceptable.

Table 4.2
Step 2: Candidate Outage Scenarios

	Scenario E1 Delta	Scenario E2 Joplin	Scenario E3 Ice Storm/Sandy	Scenario E4 Cyberattack	Scenario E5 Sandy + Cyber
Duration	12 hours	3–7 days	2 weeks	1 month	3 months
Physical effects	Base	Local	Regional	None	Regional
Cybereffects	None	None	None	Base	Regional
Power quality	Present	Not present	Not present	Present	Present
Scenario narrative	A lightning strike on base power line causes local fire and power quality event	High winds create large debris field on base and in surrounding community	An ice storm severely damages power lines and trips relays or a hurricane causes severe flooding and wind damage; off-base communications, landlines down	An adversary attacks information technology and backup power systems on the base and also physically targets critical nodes in the power system, cutting the power grid	Combination of Scenarios E3 and E4, in which an adversary launches a targeted cyberattack following or in the midst of a Sandy-like disaster
(Sample) broken plans and assumptions	Instruments and other equipment cannot restart following event; data unavailable	Off-base support personnel and fuel service unavailable because of downed lines and debris; communications capabilities lost	Off-base support personnel and fuel service unavailable because of downed lines and debris; communications capabilities lost	Instruments and other equipment cannot restart following event; data unavailable; loss of all communications	Off-base support personnel and fuel service unavailable; loss of data access; loss of communications

SOURCE: Narayanan et al., 2017

the most. Moreover, as discussed under Steps 3a and 3b, it is important to include scenarios that the installation and mission have previously experienced (and for which they have an understanding of the nature of degraded performance) and scenarios that the installation and mission have not experienced.

Outage scenarios can occur anytime: when a mission's operational tempo reflects "normal" peacetime conditions, when major operations against a foreign adversary are under way or could be soon initiated, and when operational demands fall somewhere between these two extremes. In considering cases when operational tempos are high, additional outage scenarios could be developed that explore a wider range of attacks by a "determined adversary," such as coordinated attacks on power utilities' supervisory control and data acquisition systems, transmission lines, and Air Force cyber systems. These scenarios could lead to catastrophic failure of power systems inside and outside installation fence lines. However, as noted in earlier work of Narayanan et al. (2017), what matters with a scenario is its net effect on power loss, power quality, and duration

Figure 4.4
Diverse Coverage of Scenario Space Using the Examples in Table 4.2

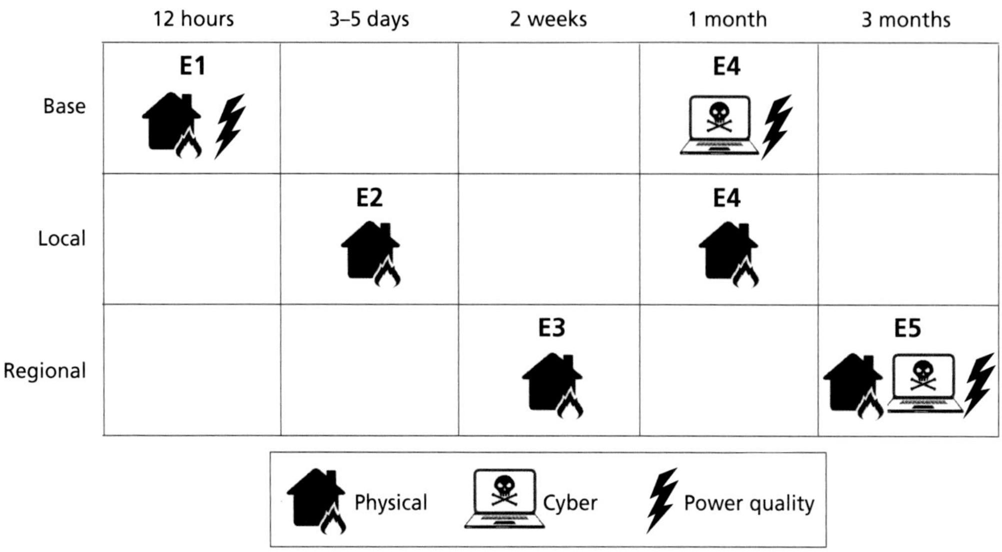

	12 hours	3–5 days	2 weeks	1 month	3 months
Base	E1 🏠⚡			E4 💻⚡	
Local		E2 🏠		E4 🏠	
Regional			E3 🏠		E5 🏠💻⚡

🏠 Physical   💻 Cyber   ⚡ Power quality

SOURCE: Narayanan et al., 2017.

Table 4.3
Structure of Event-Driven Scenarios

Condition	Description	Extent
Duration	Period over which event causes outages or other disruptions	Hours, days, weeks, months
Physical effects	Weather or terrorist events physically damage equipment or disrupt operations	Base, local, regional
Cybereffects	Internet or information technology systems are compromised or inaccessible	Base, local, regional
Power quality effects	Voltage sags or swells that damage or otherwise degrade sensitive equipment	Present, not present
Broken assumptions and plans	Critical system architecture elements break down in unexpected ways	Backup systems fail to turn on, loss of access to base, etc.

SOURCE: Narayanan et al., 2017

of the outage—regardless of the underlying cause. Hence, the approach taken in this framework remains applicable even under the most stressful scenarios.

TACC Example

In support of our running example of TACC and Scott AFB, we construct a Storm scenario, which is somewhat comparable to the Joplin scenario in Table 4.2. In this scenario, a storm causes damage to local infrastructure, creating a power outage in parts of Scott AFB, including TACC's main operations center. The local COOP site maintains power. However, the communications system shared by TACC's main building and the local COOP site goes down. This scenario serves as a reminder that power systems are often tightly integrated with other systems, such as communications. Table 4.4 defines the Storm scenario according to the dimensions in Table 4.3. Note that this scenario is for illustrative purposes only; in a full analysis of the TACC mission, one would need to examine multiple scenarios that stress the system in different ways.

Step 3a: Estimate Performance for Each Scenario Under Status Quo System Architecture

General Description

By this step, the mission owner should have in hand: an appropriate performance metric, a clearly specified performance target, and a reliable way to track and record performance (Step 1a); a significant understanding of the status quo system architecture (Step 1b); and a set of scenarios against which to test performance (Step 2). The next step is to estimate how the current system performs under different outage sce-

Table 4.4
Outage Scenario Caused by an Intense Storm with High Winds and Rain

Condition	Storm
Duration	Eight days
Physical effects	Base
Cybereffects	None
Power quality	None
Scenario narrative	A storm causes a power outage for part of the base and damages some on-base infrastructure.
(Sample) broken plans and assumptions	Communications capabilities are also degraded. Systems and power are slowly restored over time, with some issues continuing even after grid power restored.

narios. Performance should be estimated both during the outage and during any additional time required for a return to baseline levels.

For scenarios that have been previously experienced, completing Step 3a would involve reviewing any available performance data before, during, and after the outage. For scenarios not previously experienced, discussion-based or operations-based exercises will be needed to estimate performance. Discussion-based exercises should include individuals knowledgeable about the COOP plan for the mission and other contingencies that might come into play. Operations-based exercises can include red teams, drills, or full-scale physical exercises.

It may be the case that there is no significant performance degradation associated with a scenario. However, this does not mean that there is no problem; performance reflects any existing power resilience measures that might have been deployed to maintain performance at a high level during the outage, at some expense in time and resources. Until the *costs* associated with achieving this performance are explored, a determination cannot be made as to whether there is a problem in these cases. This becomes the focus of Step 3b.

Conversely, the performance degradation associated with the scenario might be significant. In this case too, one should be cautious about attributing the performance degradation to the adverse event, as other factors may have come into play. The performance metric selected in Step 1a could have a high variance, even when an organization is not dealing with an outage; thus, care should be taken in Step 3a to examine the all the factors that could influence performance during an outage. Institutional knowledge coupled with a positive record of historical performance data with reasons for any dips in performance could be quite helpful in choosing the right comparisons and disentangling correlation from causation.

TACC Application

To illustrate Step 3, we assume a case where TACC has already experienced the notional Storm scenario and a case where it has not.

Case 1: Storm Has Been Experienced

Figure 4.5 shows mission delay data for a period during which we assume that TACC experienced the Storm scenario (shaded in gray). Because this scenario has already occurred, Step 3a involves simply reviewing performance data tracked in GDSS.[6] TACC could use GDSS data to predict the effects of a future Storm scenario with the status quo architecture, fulfilling the goal of Step 3a. In this example, the performance degradation associated with a future Storm scenario looks to be fairly minimal, but

---

[6]   This, of course, would not be an option if the scenario had somehow rendered GDSS unusable. For the purposes of this discussion, we assume that either GDSS or some substitute tracking system stayed functional through the course of the outage.

Figure 4.5
Measuring Performance During Previously Experienced Scenario

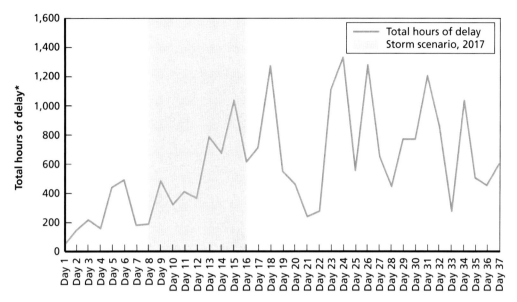

* = Delay assigned to date of takeoff.

SOURCE: Authors' analysis of GDSS data supplied by TACC.

NOTE: Gray shading shows period during which it is assumed the Storm scenario was experienced.

more detailed examinations and situational context should be considered, as discussed earlier.

For TACC, day-to-day variance in delays is completely normal due to such factors as aircraft availability, global weather, and other idiosyncratic sources of delays. For example, the weeks before the event occurred may have had fewer delays simply due to TACC receiving fewer requests. Fortunately, TACC keeps detailed records on the cause of each delay. Figure 4.6 shows the six main causes of delays during the period in which we assume the storm occurred. The data show that "logistics" and "miscellaneous" delays spiked during the event, but neither delay series is unusually high compared with its normal variance over longer periods of time.

Case 2: Storm Has Not Been Experienced

If the scenario has not been experienced previously, there is still value in reviewing any available performance data pertaining to related scenarios. In the TACC example, having a significant understanding of common causes of historical mission delays can help to anticipate future performance. However, there is no substitute for conducting a discussion-based or operational exercise aimed at gaining a better understanding of the ways in which the specific scenario is likely to affect mission performance. An operational exercise could either simulate an outage, without turning off power, or actually turn off power for a predetermined duration. If power is actually cut, then the exercise

Figure 4.6
Total Hours of Delay, by Cause

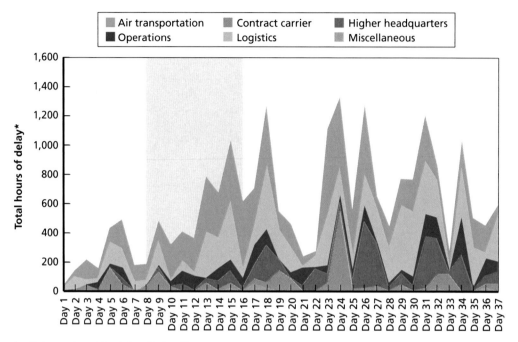

* = Delay assigned to date of takeoff.

SOURCE:Authors' analysis of GDSS data supplied by TACC.
NOTE: The gray shading shows the period in which it is assumed the storm scenario was experienced.

will need to be planned out very carefully and should be coordinated with other installations that will need to carry out mission functions that would normally be handled
at the exercise location. Additionally, backup power options and plans should be tested
before the exercise is conducted.

Step 3b: Estimate Costs for Each Scenario Under Status Quo Power Architecture
General Description

As noted earlier, the performance identified in Step 3a comes at a cost, even—and
perhaps especially—when performance does not appear to degrade during the outage
scenario. This is because of the procedures and capabilities that come into play during
the scenario. The goal of Step 3b is to measure two values under these conditions when
no additional power resilience options are considered: (1) the average annual cost of
all power-related expenses and (2) the one-time costs of operating through a given scenario (e.g., running backup generators, replacing damaged equipment) costs.

 As with Step 3a, there are two cases to consider: (1) the scenario has been experienced and performance and other relevant data exist and (2) the scenario has not
been previously experienced. For scenarios that have already been experienced, Step

3b requires gathering as much information as possible about costs incurred during the scenario. The cost categories of interest are the same in both cases.

One cost category is the average annual cost of electricity, including the amortized cost of large purchases, such as generators, across time. Step 1b should have provided significant insight about the average annual costs of power-related expenses, as these are typically incurred during standard performance and are well recorded. A second category is one-time costs that are expected to be incurred during a scenario given the status quo architecture, such as the costs of running backup power generation sources or the cost of moving operations to a COOP site, if one exists.[7] These should be known or assessed as part of Step 1b.

One-time costs can be further broken down into two categories: "recovery costs" and "heroics costs." We define *recovery costs* as those involved in fixing the source of the problem. For example, if an UPS system fails, there is a cost associated with fixing the system or purchasing a replacement. Recovery costs also include costs associated with replacing any equipment that is degraded or destroyed as a consequence of the power-related event (e.g., computers that are damaged due to an UPS system failure). We define *heroics costs* as those associated with mission and CE personnel doing more than they would under normal conditions to keep the mission going. These efforts are often creative and impressive, and lessen the effects to the mission that would have otherwise occurred if not for those going above and beyond the call of duty to "make it work." However, there are also opportunity costs in having personnel diverted from other tasks they could be performing as they cope with the demands of mitigating the effects of the outage, and there may be costs associated with unexpected overtime hours. Depending on how the event is managed, morale may decline, which could have implications for productivity post-event.

Finally, there might be costs that go beyond the officially planned mitigation response. Such costs, while they cannot be easily anticipated, should be carefully tracked as they are incurred.

The goal of Step 3b is to gather all relevant cost information for use in Equation 4.1 that defines cost as a function of scenario frequency:

Total Annualized Cost =
(Average Annualized Cost of All Power-Related Expenses) +   (Eq 4.1)
(One-Time Cost of Scenario Occurring) × (Scenario Frequency per Year).

Note that there is no need to define or consider scenario frequency at this point. Scenario frequency will enter in Step 6.

---

[7]   Not every mission can be operated elsewhere in the event that the primary operating site is degraded. Further, COOP sites are at times co-located with the primary site, which means that the same scenario affecting the primary site could affect the COOP site.

TACC Example

Using a recent power-related event TACC experienced as an example, we gained insight into the extent to which TACC tracks the different types of costs described in the previous section. Annual costs associated with buying grid power and standard maintenance of backup power resources are well known. Additionally, TACC staff who had been intimately involved in resolving the issues related to the recent disruption were able to provide estimates of a subset of recovery costs (e.g., the costs of relocating to a COOP site). We suspect we did not identify *all* recovery costs because we had a more difficult time obtaining any records or data documenting heroics costs, though we did hear anecdotally of personnel working overtime. We do not present these costs this report due to the sensitive nature of these data.

Step 4: Identify Candidate Power Resilience Options
General Description

The next step involves identifying and vetting a list of power resilience options to be analyzed. To be considered as a candidate power resilience option, the option should reduce the performance degradation resulting from or expected to result from exposure to the scenario and/or reduce the costs incurred during grid outages. Power resilience options may include materiel changes to the base system architecture (e.g., installing new and varied power generation sources) or nonmateriel changes (e.g., updates to procedures or policy). Not all power resilience options will resolve performance gaps in the same way, to the same extent, and for the same cost. In addition to differences in upfront costs, there may still be heroics and recovery costs incurred even with a power resilience option in place.

Mission owners and BCEs, harnessing their respective expertise and perspectives, can collectively come up with an initial list of viable candidate options. This list can then be vetted by an impartial subject-matter expert with access to the right tools. For example, Air Force personnel within OEA and AFCEC are likely to have relevant technological and engineering expertise, which should be leveraged when appropriate. While it does not capture costs involved in implementing nonmateriel solutions, a potentially useful resource for comparing various system architectures is the MIT Lincoln Labs resilience framework for overhauls of system architectures (Judson et al., 2016). Finally, Appendix C of Narayanan et al. (2017) provides some preliminary guidance on identifying appropriate power resilience options for different types of performance gaps. It is important to note that the analysis framework described in this chapter is designed for making decisions about different outage power resilience options and not for determining whether to pursue routine maintenance procedures. Although resource constraints always mean careful thought is required for determining what investments are appropriate, basic maintenance is critical to ensuring that systems are operating as intended. Without standard maintenance, the likelihood of unexpected system failure can significantly increase.

TACC Example

To illustrate the types of power resilience options that might be considered in Step 4, consider three notional power resilience options to improve TACC performance during the Storm scenario. Power resilience Option A calls for upgrading TACC's communications infrastructure to improve resilience to power outages. At present, certain attributes of the local COOP site's communications network limit its usefulness in the event of a base-wide outage. Having the local COOP site on a separate system would improve resilience. Power resilience Option B calls for changing the remote COOP site to ensure that TACC would have easy access to COOP resources in a wider variety of potential scenarios. Power resilience Option C calls for installing an alternative, dedicated power source (that is not a diesel generator) for TACC that has the ability to operate in grid-disconnected mode during a grid outage. Note that this list of options is strictly illustrative and is not intended to be comprehensive; it merely indicates the types of options that could be identified through discussions with TACC mission owners and Scott AFB BCEs.

Step 5a: Estimate Performance Associated with Power Resilience Options
General Description
As described previously, candidate power resilience options should reduce a performance gap for a given scenario and/or reduce the one-time costs resulting from the scenario. In Step 5, we estimate the degree to which each power resilience option meets these objectives: Step 5a examines performance improvements; Step 5b examines cost reductions.

For each power resilience option, Step 5a involves assessing the expected performance, as measured by the performance metric defined in Step 1a, if the power resilience option were in place. This information could be qualitative, based on an understanding of the expected performance for each power resilience option in a given scenario, or could result from actual experience or mathematical simulation. While a simplifying assumption would be that all power resilience options considered for a particular scenario perform equally well, in reality, options are likely to vary in effectiveness. The framework accommodates both cases.

TACC Example
In the Storm scenario, we assume that the availability of power varies across the base, with the local COOP site retaining power. In this case, a separate communications system for the local COOP site (Option A) would mean it retains both power and communications systems when the main operations center experiences an outage. Therefore, Option A could improve mission performance in the Storm scenario. However, if the entire base is without power (as could happen in other scenarios), then a separate communications system for the local COOP site is unlikely to diminish any performance gap.

Option B entails changing the location of a remote COOP site. The decision to operate out of a COOP site implies that the mission is unable (or will become unable) to meet some performance threshold with local resources. Consequently, the specific scenario conditions are not important for estimating any performance improvement that may result from execution of this power resilience option. This option should only be considered if the current remote COOP site restricts performance when operating under ideal conditions, or if the current remote COOP site has some risk of restricting performance at the times when its use is required (e.g., inclement weather, shared space).

Option C would switch TACC's power source to an independent "islanded" power generation source that only TACC could use when the rest of the base is experiencing an outage. This power resilience option would likely protect TACC's ability to continue operating as normal under a wide range of scenarios.

Step 5b: Estimate Costs Associated with Power Resilience Options
General Description

The second reason for investing in a power resilience option is to reduce the costs associated with the scenario. Similar to Step 3b, the goal of this step is to measure two values for each power resilience option: (1) the average annualized cost of all energy expenses, including any costs involved in acquiring the power resilience option (if it is a materiel solution), and (2) the one-time cost of the scenario occurring. The goal is to have these two values ready for use in Equation 4.1. Again, there is no need to define or consider scenario frequency at this point. Scenario frequency will enter in Step 6.

Each power resilience option will have a set of fixed costs, regardless of how often a scenario occurs, including the cost of installing, creating, enlisting, and maintaining new and existing base architecture, policies, and personnel. Power resilience options involving the installation of new equipment may have large fixed costs but minimal costs associated with day-to-day usage, while increases in personnel may have no large upfront costs but present a steady level of day-to-day costs. Power resilience options might increase the average annual cost of power-related expenses, such as installing and maintaining additional power infrastructure, and increasing current installation and maintenance spending. However, it is possible that a power resilience option could reduce these costs if it also reduces or offsets the cost of purchasing grid power during normal performance. Again, this cost information should be converted to an average annual cost of amortized expenses. That is, if the costs are spread smoothly across the system's lifetime, this would be the cost per year to create and maintain the new system.

In Step 3b, we described several cost categories. To the extent possible, Step 5b should assess these costs for each power resilience option for each scenario. Once Steps 5a and 5b are complete, Table 4.5 can be completed using Equation 4.1.

Table 4.5
Performance and Cost of Each Power Resilience Option for a Given Scenario

	Performance during scenario	Upfront purchase cost	Cost per scenario	Annualized cost (over option life cycle) for wide range of scenario frequencies				
				Infrequent		...		Frequent
Status quo power architecture								
Option A								
Option B								
Option C								

TACC Example

We were not able to gather detailed data on the expected costs of the Storm scenario under the three notional power resilience options due to some of the same data limitations discussed in Step 3b. That said, we are able to identify some key cost drivers for each power resilience option.

For Option A, the main fixed cost is the cost of installing a separate communications system for the local COOP site. This cost would need to be amortized across the life of the system, and an estimate of maintenance costs should also be included. Under the assumption that this would fully resolve the performance degradation associated with the Storm scenario, the costs associated with travel to the remote COOP site would not be incurred. There may potentially be relatively minor costs associated with transitioning to the local COOP site.

For Option B, there might not be a large installation cost if the new COOP site uses existing space, but establishing a new remote COOP site does likely entail some expenses in terms of staff time spent creating the plan and any resources involved in MAJCOM or higher-level approval of the establishment of a new COOP site. There would still be travel costs associated with reaching this remote COOP site, and those may be larger or smaller than the costs of reaching the previous COOP site.

For Option C, there is again an upfront cost of installation, as well as a steady stream of maintenance costs. If the power source is used outside of grid outages, this may change TACC's day-to-day cost of power. Because TACC would be able to continue operations in its main building uninterrupted, costs associated with travel to any COOP site are eliminated.

Given that the above power resilience options are highly notional, the remainder of this discussion does not reference the TACC example specifically, but rather describes how the analysis would be carried out in any context. Assuming that a full analysis of the TACC mission were to be carried out through Step 5, we observe no fundamental barriers to the Air Force and TACC leadership being able to complete Steps 6 and 7.

Step 6: Determine a Preferred Power Resilience Option for a Given Scenario

The choice between the various power resilience options and the status quo power architecture is essentially a trade-off between potential combinations of cost and performance. Options that offer lower performance at a higher cost can be easily eliminated; options that improve performance and lower costs are most desirable. The reality is that most choices will fall between these bounds, and decisionmakers will face a trade-off between cost and performance. The purpose of this framework is not to make the trade-off decision for the decisionmaker, but rather to present the option as clearly as possible to inform the decisionmaker's grasp of the choice and facilitate the value judgment.

As noted in Step 5b, the cost per year associated with the power resilience option or the status quo power architecture depends on the frequency of the scenario. However, the probabilities of occurrence of the type of rare events most likely to disrupt Air Force operations are generally unknown or have no evidentiary base. Rather than assuming the scenario frequency at the outset (and potentially getting it wrong), this framework uses an alternative approach. We ask how the preferences for options might change across a range of possible frequencies of each the scenarios.

We begin with a simple notional case for a given scenario in which each of the options is estimated to yield the same performance improvement per scenario event, as shown in the top row of Table 4.6. Therefore, options are distinguished only by their annualized costs, which include both upfront and recurring costs. Starting with the third row of Table 4.6, options are ranked for each assumed frequency. The status quo power architecture would be the preferred option for this particular scenario if the frequency of the scenario were assumed to be no greater than once annually. However, in this notional example, Option B emerges as the better option if the scenario were to occur more frequently.

Because the frequency of the scenario occurring is likely unknown, the mission owner instead needs some insight into the sensitivity of his or her decision to varying assumptions about cost. A graph like Figure 4.7 could be constructed that would show the sensitivity of the total annualized cost of each preferred option to different frequencies of the scenario. If the mission owner were indeed cost-sensitive, then the relationship between the assumed frequency of the scenario and the shape of the curve would provide useful information about the robustness of his or her choice. A notable increase in annualized cost as the frequency of the scenario increases would suggest that a mission owner—if planning for only this single scenario—would choose Option B over the status quo as the more prudent option.

Table 4.6
Notional Example of Cost and Performance Trade-Off Analysis When Options Yield the Same Performance Benefits for a Given Scenario

	Status Quo Power Architecture	Option A: Upgrade Communications Infrastructure	Option B: Change the Remote COOP Site	Option C: Add Island-Mode Enabled Power Source for Main Operations Center	Preferred Option
Performance improvement (reduced mission delays in hours) per scenario event	2	2	2	2	No preference based just on performance
Total annual performance improvement (reduced mission delays in hours)					
Annual frequency is 1X	2	2	2	2	
Annual frequency is 2X	4	4	4	4	
Annual frequency is 5X	10	10	10	10	No preference based just on performance
Annual frequency is 8X	16	16	16	16	
Annual frequency is 10X	20	20	20	20	
Total annualized cost (times $100,000)					Based on cost only:
Annual frequency is 1X	20	30	35	23	Status quo
Annual frequency is 2X	40	40	40	32	C
Annual frequency is 5X	100	70	55	59	B
Annual frequency is 8X	160	100	70	86	B
Annual frequency is 10X	200	120	80	104	B

Figure 4.7
Notional Relationship Between Total Annualized Cost and Frequency of Scenario

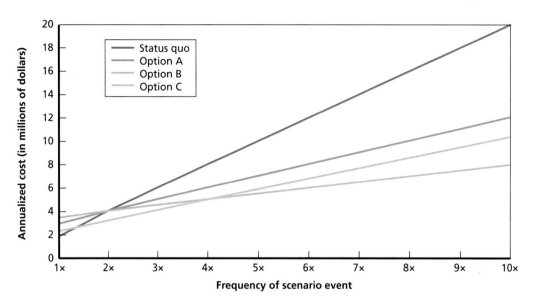

Alternatively, a mission owner might conclude that he or she has zero tolerance for the consequences of even one occurrence of the given scenario. Under these circumstances, neither cost nor per event performance improvement would be a discriminator in choosing among options: Any option would suffice, but Option B would be the more conservative choice from a budget perspective.

The information in Table 4.6 and Figure 4.7 also can be summarized in table form, as shown in Table 4.7. The left side of Table 4.7 shows a notional set of rankings that vary across five scenario frequencies. It may be possible to discern a ranking that is robust across this range of uncertainty—for this scenario—as shown in the last column of Table 4.7. This might not always be the case. Under such circumstances, the decisionmaker will need to choose another criterion (e.g., budget constraints, ease of implementation, timeliness) as a discriminator.

Unlike the example in Tables 4.6 and 4.7, in which each option was assumed to produce the same level of improvement, the robust ranking of options for a given scenario is a more difficult task when the performance improvement across the options varies. Under this more challenging case, the results might look like the notional example in Table 4.8. Option B remains as the low-cost option if the scenario were to occur more frequently. However, Option A produces the greatest total improvement in performance across all scenario frequencies when it is assumed that performance gains accrue each time the scenario event occurs.

If the mission owner has very low tolerance for outages of any duration and thus requires the highest level of performance improvement, then Option A will always be

Table 4.7
Robust Ranking of Options for a Given Scenario, Assuming Equal Performance Across Power Resilience Options

	Annual Frequency 1X	Annual Frequency 2X	Annual Frequency 5X	Annual Frequency 8X	Annual Frequency 10X	Robust Ranking of Options for This Scenario
Most preferred option	Status quo	Option C	Option B	Option B	Option B	Option B
Second-most preferred option	Option C	Status quo, Option A, or Option B	Option C	Option C	Option C	Option C
Third-most preferred option	Option A	—	Option A	Option A	Option A	Option A
Fourth-most preferred option	Option B	—	Status quo	Status quo	Status quo	Status quo

the preferred option for this particular scenario. If the mission owner has higher tolerance for outages, then the performance versus cost trade-off becomes relevant, and a graph like Figure 4.8 would need to be constructed for each frequency for this particular scenario. The most desirable region of the graph is the lower right corner, where performance is highest and cost is lowest. For a scenario frequency of once annually in the example in Figure 4.8, Options B and C would not be desirable relative to the status quo and Option A: They are more expensive and less effective. However, the choice between the status quo and Option A is less obvious if considering a scenario frequency of only once annually. In terms of cost-effectiveness (or "more bang for the buck"), both deliver one hour of reduced mission delay at a cost of $500,000 per hour, but Option A has the capacity to deliver six hours of improvement, compared with the four hours of improvement that can be delivered by the status quo over the course of a year. Missions may vary in their demand for improved performance and their willingness to pay extra for it.

For this step of the framework, the important point is that missions should be aware of the trade-off between cost and performance and its dependency on the choice of scenario and assumptions made about the frequency at which the scenario might occur.[8] When the frequency of the scenario is unknown, which would be the usual situation for all but routine maintenance–related outages, then the mission owner needs

---

[8]   While we propose a relatively simple approach to considering power resilience options, a more formal method known as 'break-even analysis' has been developed to deal with comparable public investment and regulatory problems in which the probability of an adverse outcome is unknown or unknowable. See LaTourrette and Willis (2007) for an application of break-even analysis to a Department of Homeland Security program.

Table 4.8
Notional Example of Cost and Performance Trade-Off Analysis When Options Yield Different Performance Benefits for a Given Scenario

	Status Quo Power Architecture	Option A: Upgrade Communications Infrastructure	Option B: Change the Remote COOP Site	Option C: Add Island-Mode Enabled Power Source for Main Operations Center	Preferred Option
Performance improvement (reduced mission delays in hours) per scenario event	4	6	2	2	Based on per event performance only, Option A is preferred
Total annual performance improvement (reduced mission delays in hours)					
Annual frequency is 1X	4	6	2	2	
Annual frequency is 2X	8	12	4	4	
Annual frequency is 5X	20	30	10	10	Based on performance only, Option A is preferred
Annual frequency is 8X	32	48	16	16	
Annual frequency is 10X	40	60	20	20	
Total annualized cost (times $100,000)					Based on cost only:
Annual frequency is 1X	20	30	35	23	Status quo
Annual frequency is 2X	40	40	40	32	C
Annual frequency is 5X	100	70	55	59	B
Annual frequency is 8X	160	100	70	86	B
Annual frequency is 10X	200	120	80	104	B

Table 4.8—Continued

	Status Quo Power Architecture	Option A: Upgrade Communications Infrastructure	Option B: Change the Remote COOP Site	Option C: Add Island-Mode Enabled Power Source for Main Operations Center	Preferred Option
Cost (times $100,000) of reducing one hour of mission delay					Based on both performance and cost:
Annual frequency is 1X	5	5	18	12	A or status quo
Annual frequency is 2X	5	3	10	8	A
Annual frequency is 5X	5	2	6	6	A
Annual frequency is 8X	5	2	4	5	A
Annual frequency is 10X	5	2	4	5	A

Figure 4.8
Notional Trade-Off Between Cost and Performance for a Given Scenario and a Scenario Frequency of Once Annually

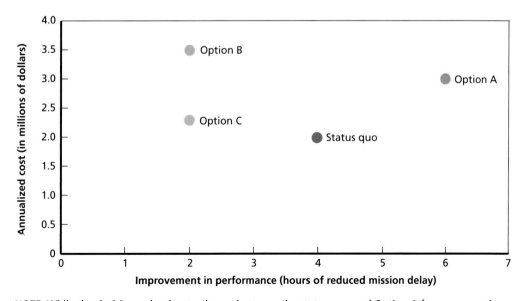

NOTE: While the decisionmaker has to choose between the status quo and Option A for an assumed scenario frequency of once annually, in this example, Option A ends up being the robust option across scenario frequencies as shown in Table 4.9.

to consider the types of trade-offs illustrated in Figure 4.8 across a range of frequencies. As with the equal-performance case described earlier, it may be possible to discern a ranking of options that is robust across a range of frequencies as shown in the last column Table 4.9. If a robust option does not emerge, the decisionmaker will need to use other criteria to select a preferred option for the scenario.

Step 7: Select Power Resilience Option That Performs Well Across Scenarios for the Mission

Decisionmakers should consider a range of scenarios in planning efforts to identify a robust option, one that will perform well across the range of scenarios considered. There may be a clear winner across the scenarios, but it is also possible that the best option to reduce costs and improve performance could be different for each scenario. In this case, if the decisionmaker does not look at performance and costs of options across the range of scenarios, this could lead to selection of one power resilience option that is less robust than another.

To identify a robust option across scenarios, Steps 3 through 6 should be repeated for each scenario considered, and the decisionmaker should choose the option that is preferred the most across the range of scenarios. Table 4.10 shows notional values resulting from a hypothetical example. In this example, Option A appears to perform best across the full range of scenarios and should be the option chosen by the decisionmaker.

A clearly preferred option may not always emerge. One way to handle this outcome is to assess the resource-intensiveness of improving the attractiveness of one or

Table 4.9
Robust Ranking of Options for a Given Scenario, Assuming Varying Performance Across Power Resilience Options

	Annual Frequency 1X	Annual Frequency 2X	Annual Frequency 5X	Annual Frequency 8X	Annual Frequency 10X	Robust Ranking of Options for This Scenario
Most preferred option	Status quo or A	Option A	Option A	Option A	Option A	Option A
Second-most preferred option	Option C	Status quo	Status quo	Option B	Option B	Status quo or Option B
Third-most preferred option	Option B	Option C	Option B or Option C	Status quo or Option C	Status Quo or Option C	Option C
Fourth-most preferred option	—	Option B	—	—	—	—

Table 4.10
Comparison of Preferences Across Scenarios

	Scenario 1	Scenario 2	Scenario 3	Scenario 4
Most preferred option	B	A or B	A	A
Second-most preferred option	C	C	B or C	B
Third-most preferred option	A	Status quo	Status quo	Status quo
Fourth-most preferred option	Status quo	—	—	C

more options. It might be the case that dedicating some additional resources to one option would boost its performance across enough scenarios to make it a robust solution for a given mission. Alternatively, one scenario may be deemed to present such a high risk to mission success that the Air Force may decide to make the investment that is likely to be most effective in improving performance under this scenario.

Data Needs and Availability as Assessed Through TACC Example

As noted throughout this chapter, not all data required to implement the proposed framework may be readily accessible. To illustrate this point, Table 4.11 presents a summary of data we were able to obtain for the TACC example, and some data challenges we faced, organized by framework step. We acknowledge that our lack of access to some of these data and information does not necessarily imply that they do not exist. Furthermore, not all missions and installations are the same, so what is true for TACC might not be true for others.

Given the importance of data to execute the proposed framework, our first recommendation (as detailed in Chapter Five) is for the Air Force to address the data collection challenges noted earlier.

Expanding the Analysis Across Multiple Missions

The framework described in this chapter addresses the problem of prioritizing power resilience investments for a single mission, across multiple installations, stressed across a range of outage scenarios. An even more challenging problem for the Air Force is to prioritize investment across multiple missions and installations. Indeed, trying to set

Table 4.11
Assessment of Data Availability Based on TACC Exercise

Framework Step	Data	Data Collected and Data Challenges
Step 1a	Historical performance data, if using as targeted performance	Data exist to construct some performance metrics, although these data may not observe all categories of mission delays.
Step 1b	Annual cost of purchasing electricity from the commercial grid	Utility bills can be used to assess the installation-wide annual cost of purchasing grid power, but it may be difficult to break up costs at the individual mission level.
	Cost of backup generation acquisition and maintenance	Data on the number of generators, both for the base as a whole and the number devoted to individual missions, along with data on acquisition costs, are available and could be used to construct amortized acquisition costs. Data on the costs of fuel and contractor maintenance should also be available. The fraction of time spent on generator maintenance can likely be estimated by those personnel (if it is not explicitly recorded).
Step 2	Define scenarios	This step does not involve explicit collection of data; it requires the identification of appropriate scenarios based on needs, risks, and situational context. Identifying a relevant scenario was feasible in the TACC exercise.
Step 3a	Scenario performance under status quo power architecture	GDSS data provided a partial measure of performance and included additional useful descriptions of the causes of delays. Anecdotal evidence about delays associated with real events may provide additional guidance. A similar tracking system might not be available for every mission.
Step 3b	Scenario costs under status quo power architecture	For scenarios that have already been experienced, information on the costs of repairing damaged infrastructure likely exists but may be divided across multiple stakeholders. If COOP plans have been activated or rehearsed, records about the travel costs associated with transitioning to and from the COOP location should be available. Records or data documenting heroics costs are likely more difficult to obtain because the associated efforts are often informal and the costs may be unreported by personnel.
Step 4	Identify candidate power resilience options	This step does not involve explicit collection of data; it requires the identification of appropriate power resilience options based on needs, risks, and situational context. Resources might be available within the Air Force (e.g., OEA) to identify candidate power resilience options for known power resilience needs. Base- and mission-level expertise is essential to developing an appropriate and usable set of candidate options.

Table 4.11—Continued

Framework Step	Data	Data Collected and Data Challenges
Step 5a	Scenario performance under power resilience options	There are likely no explicit data or models of performance outcomes in the scenarios of interest when various power resilience options are in place. The appropriate collection of stakeholders should be able to form reasonable estimates of expected performance outcomes in this context.
Step 5b	Scenario costs under power resilience options	There are likely no data or models that can readily provide the necessary cost estimates for different power resilience options. Costs are likely to be mission and location specific. Resources might be available within the Air Force (e.g., OEA) to estimate scenario costs for different power resilience options.
Step 6	Determine a preferred power resilience option for a given scenario	Completing this step does not require additional cost or performance data beyond that identified in prior steps, but it does require eliciting stakeholder preferences. The framework describes how this elicitation would proceed, provided the data required in prior steps are available.
Step 7	Select power resilience option that performs well across scenarios for the mission	As written, this step does not require additional data. It is possible that stakeholders might wish to place more weight on particular scenarios, and if so those weights would need to be elicited.

priorities across missions is not a technical choice, but a value judgement that must be made by senior leadership within the Air Force. There is no single or simple answer as to how the Air Force should prioritize power resilience–related investments across missions.

Having said this, certain attributes of missions and resilience options could be useful for prioritizing investments across missions. For instance, a mission that has no tolerance for any disruption might be prioritized over one that has some tolerance for less than full power supply for some period of time. Similarly, a resilience option that supports a whole installation, and thus produces collateral and broader benefits, might be prioritized over one that supports a single mission. However, it is possible that the single-mission solutions could be of such high priority that the consideration of broader benefits becomes secondary. This dilemma opens the door to options for cost-sharing between installation support organization funding streams and mission-specific funding streams, as discussed further in Chapter Five.

In Chapter Five, we summarize our findings and make recommendations on issues concerning the implementation of the proposed framework for valuation of power resilience investments and its place in Air Force funding and other relevant processes.

Conclusions and Recommendations

In this final chapter, we summarize our conclusions and recommend next steps for the Air Force. We begin with a brief discussion of how the Air Force can improve data collection, which is a key enabler of the valuation framework described in Chapter Four. We then go beyond the framework to address how the Air Force can improve communications between BCEs and mission owners, the linkage between asset management practices and budgeting, the alignment of organizational incentives for investment, and the incorporation of power resilience into acquisition and basing processes. These recommendations are in order of their degree of difficulty in implementation. The more challenging recommendations would require extensive involvement of other Air Force components.

Improve Data Collection

Implementing a systematic valuation framework of the type presented in this report requires the Air Force to collect and maintain different types of cost and performance data. As described in Chapter Four and summarized in Table 4.11, our examination of TACC suggested several gaps in available data that would stand in the way of the proposed framework's implementation. Cognizant of the burden any data collection effort imposes on organizations, we nonetheless find that data that can be easily gathered and compiled could pay dividends by improving the quality of the analysis embedded in the valuation framework. Several of the data gaps identified could be easily filled by recording information in one place that is spread among many within the enterprise. For example, the BCE (with support from mission owners on an installation) could be in charge of tracking, in a central location, all costs that are incurred during a disruption scenario, irrespective of the type of cost (e.g., recovery, heroics) or who bore it. Similarly, mission owners should track as much performance data as possible, and share these data with BCEs. To systematically value electric power resilience across the enterprise, mission owners and BCEs need to work together to gather, maintain, and share data.

Improve Communications Between Base Civil Engineers and Mission Owners

Increase Base Civil Engineers' View into Mission Owners' Contingency Plans

Mission owners are in the best position to know (or determine) the level of power resilience required to support their mission, whether across bases, within a base across facilities or infrastructure assets, or for individual assets. BCEs, given their association with particular installations and physical infrastructure, tend to take a perspective tied to their locations and to material solutions to power resilience problems. If mission owners can support their missions by utilizing another base or some alternate means on-base (e.g., through the exercise of a COOP plan), that could reduce the power resilience demands that BCEs might otherwise attempt to meet.

Our recommendation to enhance communications between BCEs and mission owners is similar to the Air Force's effort over several decades to push operators and logisticians to better communicate with one another in contingency planning and execution (see, for example, Leftwich et al., 2002). The intent in those contexts was to encourage operators to share more information about their plans in advance of operations so that logisticians could plan for resources accordingly, and develop creative solutions and alternatives as needed. Akin to this longstanding and ongoing effort for logistics, our recommendation would require BCEs to learn to "speak ops" and understand how operators think about and plan for mission assurance.

Another dimension of communications relates to collaboration about the breadth and characteristics of planning scenarios for disruptive outages that the IC and mission owners should jointly consider. Recent Air Force guidance requiring sufficient backup power for mission-critical infrastructure for at least seven days or the time needed to relocate the mission, whichever is longer, is a step in the right direction (Air Force Policy Directive 90-17, 2016). However, there is room to be even more imaginative about disruption scenarios and their many dimensions, and to inculcate scenario-based thinking into all process elements that touch power valuation in one way or another.

Implementation of this recommendation could be led by BCEs, although SAF/IE might need to specify in a policy document that mission owners share contingency plans with BCEs. Mission owner buy-in is a critical enabler of this recommendation.

Create a Forum for Evaluating Incremental Versus Transformational Power Resilience Options for a Given Mission

The valuation framework proposed in Chapter Four attempts to close a gap in project evaluation by providing a more systematic way to pick among competing power resilience options for known or anticipated performance gaps for a given mission. This approach allows a range of candidate power resilience options to be considered (Step 4), without discriminating on the basis of their size or nature. Power resilience options could include one or more small projects to replace components or assets over time; others might involve reengineering some aspect of power resilience at a base, thus requiring major construction, renovation, or modernization. The Air Force's processes, however, differentiate between FSRM, MILCON, and third-party financing as funding vehicles. As explained in Chapter Three, these funding streams differ greatly in how they value power resilience and bring proposed projects to fruition.

The separation of funding streams and differentiation of processes creates a barrier to reaching system- or mission-level trade-offs that might benefit mission owners and the Air Force through better integrated power resilience options. Today, FSRM and MILCON projects follow separate paths. The decision to propose projects in either process lies at the base level. Each funding path has its own respective bureaucratic processes and prioritization schemes.

The Air Force would benefit from having a forum in which incremental (i.e., FSRM) versus transformational (i.e., MILCON) power resilience options could be systematically and strategically weighed and decided on for a given mission. This would expand the options in the funding-stream deliberations and highlight the need for picking projects that improve power resilience regardless of which funding stream might be more readily available than another. In the longer term, these deliberations could inform a broader institutional calculus about the types of power resilience options to pursue. A side benefit of this approach would be the institutional learning that would come with project advocates developing a better understanding about which types of solutions best fit which circumstances.

AFCEC could take the lead for such a forum, which would, at least initially, need to occur outside the current FSRM and MILCON deliberation processes. Mission owners and operators would need to actively participate in weighing the trade-offs.

Link Asset Management Practices and Budgeting

Use Standard Asset Management Principles for Routine Maintenance and Repairs

This recommendation is more difficult than it might appear. While there is disagreement about the proper way to estimate the total amount of funding required to sustain Air Force infrastructure in good working condition (Mills et al., 2017), routine maintenance of infrastructure is more straightforward. In many cases, equipment manufacturers specify the routine maintenance required to sustain each component in working condition. BCEs estimate the labor hours required to support manufacturer-stated routine maintenance tasks, set minimum manpower levels to meet these routine tasks, and add manpower to support urgent and emergency tasks in response to on-base customer requests.

This recommendation calls on Air Force leadership to commit to scheduling routine maintenance and repairs of base infrastructure using standard asset management principles. While the Air Force does not have control over its overall congressional appropriation, the Air Force does have control over the case it makes to DoD, the Office of Management and Budget, and appropriators for adequate funding for routine maintenance. Routine maintenance is the first line of support for sustainment of infrastructure assets. Maintenance also is foundational to power resilience, ensuring a baseline level of reliability for energy systems, consistent with expectations at the time of acquisition. Such a commitment would need to come from the highest levels of the Air Force, such as the Chief of Staff of the Air Force, with the technical and cost specifications supported by SAF/IE or AFCEC.

An additional factor that drives asset management is the manner in which assets are classified. For example, facilities are categorized using codes that may not accurately account for the functions carried out with them. An administrative building on one base might house human resources functions but on another base house TACC. Both would be classified as administrative buildings, but their vastly different demands for assured power would not be captured in current project ranking formulas. As the only rapid global mobility operation supporting combat delivery and strategic airlift, air refueling and aeromedical evacuation operations around the world, TACC capabilities are called on 24/7. [1]

Consolidate the Budgeting and Management of Electricity, Communications, and HVAC Systems in Investment and Asset Management

Historically, budgeting and management of electricity, communications, HVAC, and control systems are handled on Air Force installations through separate organizations, contracts and subcontracts, procurements, and personnel. This division of labor made sense when these systems were largely decoupled from one another: when communications meant copper wire telephony, and before water-cooled air conditioners were mandatory equipment for protecting sensitive servers and other electronics. These conditions no longer apply. Indeed, the interdependencies among these systems are ubiquitous. As such, the separation is artificial and at times creates blind spots when identifying vulnerabilities.

To meet present and future needs for resilience in power systems, the Air Force should consider consolidating the architectural, engineering, design, and operation and maintenance services for these interrelated support systems under a single umbrella organization. The intent would be to expand the options for building resilience and efficiency into the design of these systems and their subsequent upgrades, and, in the process, streamline procurement, construction, and maintenance.

---

[1]   For organizations such as TACC that do not have a program office or formal injects into the PPBE process, getting power-related requirements filled can be challenging. TACC is a functional AOC, but it is not classified as a falconer weapon system. This means that the organization does not have the same accommodations as other AOCs, such as dedicated communications squadrons or dedicated facility space, and consequently does not rise to the priority for power resilience measures afforded to critical systems.

Cost-Sharing to Better Align Incentives for Investment

Air Force missions are highly electrified and networked across installations. What historically have been installation "support" systems are now integral to the success of most Air Force missions. Missions and installations compete for centralized funds. Under current Air Force practice, AFCEC uses a prioritization model to score and rank infrastructure projects across the Air Force enterprise. Funds for selected projects are then allocated to BCEs to cover installation-related power needs. The current AFCEC formula does not explicitly distinguish between projects that benefit just one mission and those that benefit multiple missions carried out at a given base. As a result, highly specialized and critical missions that are particularly sensitive to power outages or degradation in power quality compete for the same general infrastructure funds as less critical missions. The question is whether the current process produces an allocation of limited Air Force investment dollars that leads to the greatest overall improvements in mission resilience. As it stands, not all requests are met in any given FY, and requests can be framed in ways likely to yield higher priority on the IPL.

In general, for nonmilitary public infrastructure systems with clearly identified beneficiaries (in this case, mission owners), pooled investment resources are more efficiently allocated when the costs of providing services are internalized at least in part by beneficiaries. Put another way, beneficiaries who demonstrate their commitment to the investment by sharing some of the cost are said to have "skin in the game." This is done, for example, through congressionally mandated cost-sharing formulas, most notably by the U.S. Army Corps of Engineers in their civil works activities (Kress et al., 2016). Congress established these cost-sharing arrangements to help ration limited federal funding among states and local governments throughout the nation.

Air Force power resilience projects, by virtue of the Air Force's role in national security, are different from equivalent civilian projects. As such, there are arguments for and against implementing a cost-sharing arrangement within the Air Force for power resilience investments. In lieu of the Air Force's current, primarily administrative allocation of limited infrastructure dollars, a cost-sharing approach could be taken in which a class of mission owners with the most-critical power demands would absorb some of the costs of power resilience into their own budgets. BCEs would still have overall operational responsibility, but the allocation of investment funding would be determined in part by the value to the mission, as demonstrated by the mission owner's willingness to share in the costs. The PAF valuation framework would still apply to decide which investments to fund, but mission owners who shared in their projects' costs would assume a larger role, gaining more control of the timing and type of investment. However, for this approach to be viable, the Air Force would need an easy transactional mechanism that would enable a mission owner to share in the investment and sustainment costs with BCEs. Such a mechanism does not exist at present and therefore would need to be developed.

An alternative approach would be to maintain the existing IPL priority-setting formula but inject a manual intervention to prioritize and allocate funds to projects outside of the funding formula. This manual override is done to some extent now, when AFCEC takes the valuation model as one input, then applies institutional priorities and emphasis as needed. While it may work in some cases, it also lacks the assurance of consistency and transparency.

The decision of which approach to take would need to be made by SAF/IE, potentially with approval from the Chief of Staff of the Air Force. AFCEC would have responsibility for ensuring power resilience projects were consistent with current installation operations, manual intervention, or cost-sharing approaches, but the burden of funding would shift more toward mission owners and their internalization of the costs of power resilience appropriate to their specific needs.

Incorporate Power Resilience into Acquisition and Basing Processes

Under current practice, measures to increase the power resilience of a fielded weapon system and its supporting installation are implemented after acquisition and deployment. However, the design and intended function of a weapon system, which are known well before the system is deployed, could provide substantial insight into the possible types of power resilience needs of a system. As has been done in the field of green chemistry and manufacturing, innovations in design and processes either before or during acquisition can eliminate some unintended consequences downstream. In the context of power resilience, the aim would be to consider power resilience earlier in the process of acquiring and locating (i.e., basing) these systems.

Weapon System Acquisition

On the acquisition side, the assumption is that by addressing power resilience needs earlier in the life cycle of the system before too many design constraints are imposed, solutions could be more efficient, affordable, integrated, and creative. This approach is similar to Air Force efforts in recent years to bring consideration of sustainment costs earlier into the weapon system acquisition process (Drew, McGarvey, and Buryk, 2013).

Appendix C provides a broad overview of the Joint Capabilities Integration and Development System (JCIDS), Defense Acquisition System (DAS), and Air Force–specific processes that support the JCIDS and DAS where performance requirements are developed and modified. This recommendation could lead to two pathways for change. First, the option set for dealing with power resilience–related infrastructure problems could be confronted early, when design changes could still be made, rather than after a system has been completely fielded. Second, weapon systems and their supporting systems could be developed in tandem, with support systems potentially being redesigned at a later time to ensure greater resilience. Such a change would surely involve the Chief of Staff of the Air Force and the Secretary of the Air Force, given the gravity and scope of making changes to the acquisition process.

Strategic Basing

While basing decisions are made much later than the design of a weapon system itself, the choice of installation can have direct implications for a weapon system's vulnerabilities once it is deployed. The Air Force manages any proposed significant changes and additions to the locations of weapon systems and personnel through its strategic basing process. Some notable and recent examples of strategic basing decisions are the evaluations of installations for hosting the F-35A and KC-46A weapon systems. The strategic basing process is a multistep process that considers an array of criteria, including but not limited to an installation's ability to execute the mission, capacity to host the unit, environmental effects of the unit, and economic factors, such as locality and construction cost factors associated with individual installations. Currently, the only place where power is considered at all is under the category of capacity, i.e., the capacity of an installation's power and HVAC systems to accommodate the new weapon system. Power resilience is not explicitly considered.

As reported in a 2013 PAF report on strategic basing,

> One group of interviewees reported that environmental criteria were neither emphasized enough nor fully developed, relative to the risks that environmental aspects might introduce to successful basing actions. Environmental criteria do not evaluate a basing action within the context of the total capacity for growth at an installation, but rather whether there are issues with adding just this basing action, they stressed. The installation's available capacity for growth in water use, storm water handling, waste management, utility needs, energy production, local air pollutants, and other indicators are not evaluated. Many of these aspects are covered in installation development plans but not explicitly incorporated into the strategic basing process. (Samaras et al., 2016)

Considering an installation's vulnerabilities to a diverse set of power outage scenarios as part of the strategic basing process can help to ensure that locations that offer an appropriate level of power resilience are selected for a given weapon system. The Air Force Strategic Basing Division at HAF is the designated office for managing the strategic basing process, but its process also involves numerous actors with varying incentives, information, and resources. The Secretary of the Air Force ultimately approves the criteria for a given strategic basing decision (different criteria may apply to different situations), and the Chief of Staff of the Air Force is also highly involved. Therefore, the decision to add power resilience to the strategic basing process in a single case or more generally would need to be made by the Secretary of the Air Force with involvement by the Chief of Staff of the Air Force, with input from their respective staffs.

Concluding Thoughts

The framework and recommendations presented in this report are aimed at improving the Air Force's ability to make risk-informed decisions about power resilience investments. The absence of a structured, information-driven, and mission-centric framework for evaluating power resilience options can force decisionmakers to rely solely on their subjective judgment to make decisions regarding which power resilience investments to fund and when. While the judgment of experts and key stakeholders certainly plays a role in any investment decision (as detailed in Step 6 of the framework), this judgment should be informed by an understanding of the performance and costs of competing options across a range of scenarios. The presented framework meets this need.

The framework is not meant to be a one-stop tool than can be used to generate "1 to n" lists of prioritized investments. Rather, the goal is to identify and clearly present decision-relevant performance and cost information that a decisionmaker can use to choose preferred power resilience options for a given mission across a range of scenarios. The ultimate goal is to prioritize power resilience–related investments across Air Force missions, but starting with a single mission offers a tangible first step toward a more-generalized valuation method.

The recommendations outlined in this chapter highlight that having a valuation framework alone does not guarantee sound investment decisionmaking. Incorporating standard asset management principles and best practices; improving communications between key stakeholders and aligning their incentives; and considering power resilience in multiple processes are all essential components of a robust power resilience valuation strategy.

Literature Review of Risk Perception by Individuals

The valuation framework described in Chapter Four calls for Air Force decisionmakers to state their preferences between various power resilience options, given information about costs and performance in specific scenarios. These preferences depend on how individuals perceive risk. Therefore, we reviewed the literature from psychology, behavioral economics, and finance to understand how individuals and policymakers think about and insure against the risk of catastrophes. By *catastrophe*, we refer to any event that occurs with a very low probability but also carries tremendously high consequences. Catastrophes include unexpected and devastating natural disasters, such as floods, earthquakes, wildfires, and hurricanes, but they also include terrorist attacks, financial and macroeconomic crises, and even a major power outage for an Air Force installation. To borrow terminology from Taleb (2007), catastrophes are negative "black swans," rare but high-impact events.

We begin by summarizing evidence on how individuals perceive the risk of catastrophes, showing some evidence that individuals underestimate these risks and other evidence that they overestimate these risks. In reviewing this evidence, we explain how inaccurate perceptions of risk lead to suboptimal behavioral responses. We conclude by exploring how policy interventions may be used to help agents make more-robust decisions in the face of the risks of catastrophes.

Estimating Low-Probability Events: Availability Heuristic and Probability Neglect

A vast literature in psychology and behavioral economics suggests that people have difficulties formulating their beliefs about the likelihood of events. Because catastrophes are by definition infrequent, inferring their likelihood from previous experiences is difficult. To simplify the problem, people tend to rely on heuristics, and consequently have systematic biases when assessing and making judgements about the risks they face. In thinking about the probability of catastrophes, one important heuristic upon which individuals rely is the *availability heuristic*. If an event is easy to recall, perhaps because it happened recently or because memories of the experience are especially vivid or detailed, people assign greater subjective probability to that event. Psychologically, events that are particularly available seem to be far more numerous or likely than events that happened less recently, or are less vividly remembered, but are nevertheless equally probable (Gilovich, Griffin, and Kahneman, 2002).

The availability heuristic can explain why people systematically purchase flood insurance more frequently after (but not before) they experience a disastrous flood that ruins their personal property.[1] Conversely, it can also explain why owners of flood insurance policies frequently allow those policies to lapse after two to four years, even when they live in high-risk, flood-prone areas (Michel-Kerjan and Kousky, 2010; Michel-Kerjan, Lemoyne de Forges, and Kunreuther, 2012). Similar patterns for insuring against the risks of other natural disasters, such as fires or earthquakes, have also been observed (Slovic, 2000).[2] As Kahneman (2011) states, because of the availability heuristic, "[t]he dynamics of memory help explain the recurrent cycles of disaster, concern, and growing complacency." Furthermore, when people are not closely attuned to the probability that harm will occur, they suffer from "probability neglect," treating risks as if they were nonexistent, even though the likelihood of harm over a lifetime is far from trivial (Sunstein, 2002).

Probability Weighting: Overestimating Tail Probabilities

Although the availability heuristic and probability neglect may make people underestimate catastrophe risks, there is also evidence that in some situations, individuals instead overestimate the probability of "tail events" (Barberis, 2013). For example, in studying perceived mortality risks, Lichtenstein et al. (1978) find that people significantly overestimate the frequency of rare causes of death. As another example, when buying home or automobile insurance, many individuals choose policies with low deductibles and high premiums. Implicitly, those buyers must perceive that a low probability disaster is actually much more likely than it is, which leads them to choose policies with lower out-of-pocket expenses when those disasters take place (Barseghyan et al., 2013).

---

[1] Despite government programs to incentivize the purchase of flood insurance, a *New York Times* article published a few days after Hurricane Katrina in 2005 showed that in Louisiana parishes affected by the hurricane, the percentage of homeowners with flood insurance ranged from 58 percent in St. Bernard's to 7 percent in Tangipahoa. Six out of ten residents in Orleans Parish had no flood insurance (Bayot, 2005).

[2] In a 1974 survey of more than 1,000 California homeowners in earthquake-prone areas, researchers found that only 12 percent of respondents had adopted any protective measures (Kunreuther et al., 1978). Fifteen years later, in a 1989 survey of homeowners in four earthquake prone counties, only 5 to 9 percent of respondents reported adopting any loss reduction measures (Palm et al., 1990). Despite intensive media coverage of hurricanes and floods in 2004 and 2005, a survey of 1,100 adults living along the Atlantic and Gulf Coasts undertaken in May 2006 found that 83 percent of respondents had undertaken no attempts fortify their homes (Goodnough, 2006).

When catastrophic events, such as a terrorist attack, produce intense emotions, people focus on the adverse outcomes of these events, and they lose sight of the probability of such an episode occurring. This leads them to inflate the likelihood of those events and to make suboptimal choices.[3] For example, in the year after the events of the September 11, 2001, terrorist attacks, heightened anxiety toward flying increased the demand for driving on interstate highways. The increase in traffic volumes caused an estimated 1,500 additional deaths from traffic accidents over that period (Gigerenzer, 2006).

Policy Responses

As we have discussed, inaccurate perceptions of catastrophe risk can lead individuals to either underestimate or overestimate the likelihood of rare events. This can lead to behavioral distortions with serious consequences. While individuals who overestimate the probability of negative risks may take exaggerated precautionary responses, individuals who underestimate catastrophe risks will seem woefully underprepared when they take place.

Interestingly, when individuals and governments actually take precautions to insure against the risk of catastrophes, they typically do so in ways that are only adequate to insure against the worst disaster that has ever actually been experienced. For example, civilizations as far back as ancient Egypt have tracked the high-water mark of rivers that periodically flooded. When relocating buildings away from flood zones, or designing dams or levees to mitigate flood risks, governments and individuals typically rely on these high-water marks, never anticipating that a flood might rise higher than the existing high-water mark. Because such a disaster has never been experienced before, it is not easy for it to come to mind (Kahneman, 2011).

Because of the inherent biases individuals have in making judgements about risks, one policy recommendation for improving behavior is to do a better job of assessing risk statistically, and to provide more-accurate, objective risk assessments to decision-makers in a digestible, coherent way. To the extent that unexpected power outages on Air Force installations could be caused by natural disasters, such as flooding, hurricanes, earthquakes, or wildfires, it may be possible to make better predictions using enriched climate or environmental models. However, for a high-consequence, low-probability event such as a catastrophic power outage, the reality is that we might never know how likely those possible catastrophes are to happen, if they ever happen at all, or how consequential they might be (Michel-Kerjan and Slovic, 2010). Because these

---

[3]   Kahneman (2011) explains that terrorism is so effective because it "induces an availability cascade. An extremely vivid image of death and damage, constantly reinforced by media attention and frequent conversations, becomes highly accessible, especially if it is associated with a specific situation" (pp. 322–323).

events happen so infrequently, inferring their likelihood from historical experience is extremely difficult (Taleb, 2007).

One policy lever available is investment in insurance products; however, in many insurance markets, individuals tend to underinvest in disaster risk, and this has provided one rationale for mandating compulsory insurance (Sandroni and Squintani, 2007). For example, many states require that individuals carry some minimum automobile insurance coverage, and the federal government requires new homeowners to purchase flood insurance when buying homes in flood-prone areas. To an extent, the Air Force already invests in insurance products. In many ways, a backup power supply system is like an insurance product, which hedges an installation against the risk of a power outage.

Another approach that may be more effective than trying to estimate risks or mandating insurance is for policymakers to use scenario planning to focus on the consequences of risks (Taleb, Goldstein, and Spitznagel, 2009). Using a framework such as the one presented in this work to evaluate the possible effect of extreme scenarios for the performance of particular Air Force missions can help policymakers better plan for their eventualities. Detailed scenarios that vividly illustrate both the vulnerabilities and resilience of particular weapon systems or communications infrastructure to both short- and long-term power outages may raise awareness and lead policymakers to make more-sensible investments in energy assurance. Using scenario analyses to understand power resiliency needs can help to ensure that operational commanders and BCEs sufficiently focus on off-normal operational conditions and do not prioritize efficiency gains over all else when planning for power resilience.

Non-DoD Valuation Approaches and Processes

As noted in Chapter Two, we sought to collect lessons from how nonmilitary organizations go about valuing and prioritizing their power resilience investments. This appendix outlines the criteria we used to find relevant organizations and summarizes what we learned from six organizations: NASA, two hospital systems, a university system, an electric utility, and a regional transmission organization. Although no single organization or corresponding sector offered a direct analog to the Air Force, the organizations we examined provided some useful context for the problems the Air Force faces.

Choosing Non-DoD Organizations Based on Air Force Attributes

To guide our selection of relevant organizations, we identified a small set of organizational attributes that capture essential features of the Air Force that are most relevant to its dependence on electricity and its needs for high reliability and power quality. We proceeded on the assumption that no organization would be identical to or capture the complexities of the Air Force, but that some might share some relevant features. The attributes we settled on, in no particular order, were as follows:

- **Distributed locations:** Air Force missions are carried out at many locations throughout the United States and abroad, and thus require a high degree of networking of communications, missions, and support for such functions as logistics and personnel management.
- **Nested levels of decisionmaking:** The Air Force has a hierarchical organizational structure with investment decisions being made at all levels.
- **Diverse set of missions:** Air Force missions vary widely, from traditional flying missions to intelligence gathering missions to humanitarian aid and disaster relief support. Therefore, we sought organizations that have distinct, varied, and interrelated missions that have different power needs.
- **Formal procurement processes:** DoD (and, thus, the Air Force) uses highly structured and formalized budgeting and procurement processes.

- **Effect on lives lost:** Disruption to Air Force missions may carry serious consequences for national security and the protection of human lives.
- **Recent power disruption:** This attribute does not relate to organizational makeup, but, for purposes of this research effort, we wanted to focus on organization that may have recently changed or amended policies in response to a power disruption.

We mapped candidate types of organizations onto these attributes, as shown in Table B.1. An organization was coded as a "yes" if we could link the Air Force attribute to the organization.

Summary of Cases

In our interviews with officials in the selected organizations, we asked about approaches to energy assurance, associated processes and investment decisions, and metrics used to determine the value of assured access to power, including any risk calculations. A summary of these interviews and the results of a supplementary literature review follows, organized by type of organization.

Table B.1
Mapping of Attributes to Candidate Organizations

	Distributed Locations	Nested Levels of Decisionmaking	Diverse Set of Missions	Formal Procurement Process	Effect on Loss of Life	Recent Power Disruption
Air Force	Yes	Yes	Yes	Yes	Yes	Yes
Non-DoD federal agencies				Yes		
NASA	Yes	Yes		Yes		
University systems	Yes	Yes	Yes		Yes[a]	
Hospital systems	Yes				Yes	Yes
Power utilities				Yes		Yes
Relief operations[b]	Yes				Yes	
Commercial airlines[b]	Yes					Yes
Online retailers[b]	Yes		Yes			

[a] "Loss of life" in this example pertains to university hospitals which may be part of university systems.
[b] Organizations of this type were not interviewed.

NASA Launch Services Program

NASA is a federal organization with distributed locations, nested levels of decisionmaking, and a highly structured procurement process. Another RAND analysis identified these attributes when categorizing components of organizational risk across NASA (Gerstein et al., 2017). The RAND researchers identified "multiple levels of management involved in work," "number of locations involved in work," and "dispersed management of projects and funds," which map to our definitions of distributed locations, nested levels of decisionmaking, and diverse set of missions (carried out at many locations) (Gerstein et al., 2017).

For additional insight on energy assurance policy, we reviewed NASA-wide directives that focus on any directive that may provide additional insight on program management, budget formulation, energy risk, and facilities maintenance (NASA, undated). Energy assurance topics are predominantly within the purview of the NASA Environmental Management Division and its Energy Management Program (NASA Procedural Requirement [NPR] 8570.1A, 2013). In addition to activities associated with energy efficiency, conservation, and system maintenance, NASA policy outlines the Energy Management Program's role to manage energy risk to mission through strategy development and project planning (NPR 8570.1A, 2013; NASA Policy Directive [NPD] 8500.1C, 2013).

At the recommendation of personnel at Vandenberg AFB, we reached out to the NASA Launch Services Program (LSP) for additional information about NASA policy and practice. LSP organizes and manages all expendable load vehicle missions across six sites: Kennedy Space Center at Cape Canaveral Air Force Station in Florida; Vandenberg AFB; Wallops Flight Facility in Virginia; the Ronald Reagan Ballistic Missile Defense Test Site in the Marshall Islands; and the Pacific Spaceport Complex in Alaska (Bedell and Norman, 2015, slide 4; Mendoza-Hill, 2015). Some of these sites are owned by NASA; for others, NASA is a tenant. LSP supports missions from premission planning to launch, which can take several years and starts with Phase I, Mission Planning (Mendoza-Hill, 2015; NASA, 2012). The first phase for LSP includes defining spacecraft and facility requirements, with a goal to create estimates for the annual PPBE process (Mendoza-Hill, 2015).

The annual PPBE process includes developing the budget year as well as the next four out-years (five-year budget) (NPR 9420.1A, 2016). As in other federal agencies, the needs of missions are balanced with the resources available, which intrinsically assigns relative value and need for mission delivery through budget allowances (NPR 9420.1A, 2016). To make these determinations,

> Agency leaders consider, among other factors: budget limits and programmatic guidance from the Administration and Office of Management and Budget (OMB); recent legislation, including authorization and appropriation acts; availability and costs of institutional capabilities; outstanding issues that arose during the prior cycle of budget formulation; and proposed future investment strategies.

The [strategic programming guidance (SPG)] documents this guidance and establishes a starting point from which the Agency will formulate the budget. (NPR 9420.1A, 2016)

Programs and mission directorates undergo studies, which feed into and support PPBE decisions (NPR 9420.1A, 2016). LSP completes facility assessments every two years to ensure that systems meet mission requirements. For example, these reviews ensure that generators are up to date and provide the basis for equipment (e.g., UPS system) updates or modifications. When energy investments are required, the decision level depends on the estimated price for the investment. For investments under $1 million, the program director has some discretion on how to use programmed funds. For those above $1 million, approval must be obtained from headquarters, and for investments exceeding $10 million, the decision can take several years for approval.

As of July 2015, LSP successfully executed 79 full missions (Bedell and Norman, 2015). To ensure each of these launches is secure, LSP conducts reviews at each phase pre-launch, including those regarding energy assurance issues (power outages in particular can delay launches) (NPD 8610.24C, 2005). Due to orbital dynamics, even relatively short power outages can result in multiyear delays, which in turn require additional resources and add unplanned costs to the mission. For example, LSP estimates that more than eight hours without power requires a coordinated shutdown of essential communications equipment. Also, if the outage were to occur in the countdown phase, NASA follows a "graceful degradation" process to avoid damage to the launch vehicle, standing down the mission. These requirements are set to minimize mission impact and for public safety reasons.

Due to its stated importance, mission impact from power outages is determined by cost and length of delay to the launch, which can be defined in standard terms across launches. This contrasts the Air Force context, where the effects of delays vary across missions. Instead, prioritization of missions and potential effects on those missions are determined relatively easily.

Overall, we did not identify any hidden best practices within the decisionmaking process at NASA concerning energy assurance prioritization and investment. The budgeting process is very similar to the Air Force process. One difference for LSP is that determining priorities between launches (missions) seems relatively easy compared with determining priorities across the Air Force diverse missions. Comparing launch to launch is like comparing apples to apples; whereas, the whole of the Air Force mission set does not allow for this direct comparison. Other prevailing factors may be in play; however, this case suggests that consistent mission classification can provide a basis for translating mission requirements to power requirements and consequent setting of investment priorities.

Hospital Systems

Hospital systems, given their life-saving mission, provide a useful analog to the Air Force's mission criticality. Often, health care organizations include facilities distributed over a wide area that provide varying levels and types of health services. Additionally, a recent survey of hospital systems suggests that, for some hospital systems, power outages occur frequently.[1]

We interviewed officials from two hospital systems in different parts of the country, which we refer to as Hospital System A and Hospital System B. Our discussions focused on how the hospital system's mission to reduce suffering and save lives are affected by power disruptions and what power planning they do to reduce the effects to this critical mission. Hospital System A consists of hundreds of facilities across several states, with most electrical power demand met through grid connected commercial utilities. Hospital System B consists of more than 1,000 buildings, including hospitals, a medical school, a medical research facility, and many outpatient treatment buildings. The majority of Hospital System B facilities are connected to the commercial grid, and its main location has its own cogeneration plant, which supplies 50 percent of its peak power needs and can provide about 30 percent to 40 percent of power needs if the commercial grid power is disrupted.

For Hospital System A, decisions to investment in power resilience are driven by regulatory compliance and cost savings. For example, Hospital System A recently installed a microgrid in one location and solar panels at several other locations as cost-saving measures. The interviewee noted that although these measures may add resilience to their energy system, economic benefit drove implementation, not resilience. We also learned that requirements set by state and national regulators motivate resilience decisions instead of strategic organizational goals or guidance. For example, hospitals in certain seismic zones are required by the National Fire Protection Association (NFPA) 110, *Standard for Emergency and Standby Power Systems*, to ensure backup power for 96 hours for certain types of systems and critical services within the hospital (typically achieved through having 96 hours of fuel on hand to run emergency generators but not considering a range of hazards). Hospital System A has facilities that are subject to this standard and has decided to adopt this across all of its facilities, irrespective of the seismic zone of the facility.

Because power resilience investments are driven by regulatory compliance and cost savings, there is no process of translating mission requirements into power requirements for Hospital System A. Instead, planning processes seek to ensure that budgets accommodate what the regulations require and the decisionmakers in this system see

---

[1] Seventeen percent of surveyed health care organizations experienced six or more outages from 2011 to 2014, according to an American Society for Healthcare Engineering two-part survey (Flannery, 2014).

no incentive to go beyond compliance standards,[2] as the regulations have been adequate for their power needs.

However, recent events, such as Hurricane Sandy, have demonstrated that these regulations have not been enough to prevent widespread outages at all health care facilities (Carron, 2016; U.S. Department of Health and Human Services, Office of the Inspector General, 2014). According to the U.S. Department of Health and Human Services, Office of the Inspector General (2014), 40 percent of hospitals in the area affected by Hurricane Sandy experience electrical outages, and 16 percent could not rely on backup generators to supply reliable power.

Hospital System B happens to be one of the hospitals that could not rely on its emergency power (due to flooding), and that experience has changed the hospital system's processes.[3] Hospital System B's energy planning is now guided by a set of scenarios that stresses a broader set of assumptions than are typically used for outage scenarios. While only required to have 96 hours of backup generation, Hospital System B chooses to have enough diesel fuel at all times to last several hundred hours. Additionally, protocols have been enacted such that 48 hours into a major event, the hospital starts to evacuate. Hospital System B indicated that, on top of these emergency protocols and backup plans, very little can be done to ensure that the regional utility is sufficiently managing the grid and their systems.[4]

University System

Similar to the Air Force, a state university system has diverse mission requirements (such as for research, health services, student housing, and administrative offices) across many campus locations and with decisionmaking at both campus and system-wide levels. In case of an outage, this means that academic and administrative priorities may need to be balanced against priorities of a hospital or research facilities. We interviewed the energy manager for one campus of a state system with nine campuses. Our focus in this interview was on understanding planning processes at each level within the university; the interface between campus planning and systemwide planning, funding, evaluation and prioritization of projects; and any examples of power disruptions.

For context, this campus has a microgrid and cogeneration plant that connect to the regional utility at one location. Approximately 90 percent of campus demand

---

[2] Hospitals operate in a highly regulated environment. For example, the NFPA publishes several codes and standards for electrical systems, including NFPA 70, *National Electrical Code*, NFPA 99, *Healthcare Facilities Code*, and NFPA 110, *Standard for Emergency and Standby Power Systems* (Carron, 2016). The Joint Commission and state regulators may also drive power requirements.

[3] According to U.S. Department of Health and Human Services, Office of the Inspector General (2014), several hospitals reported that emergency planning had been valuable during the storm and revisions to their plans were conducted after learning from the experience.

[4] The interviewee also indicated that the regional utility is a top-performer in this regard.

is met through onsite generation and is under the control of the facilities department. Hurricane Sandy demonstrated the need for robust planning in the case of high-consequence events, and this campus now conducts a strategic planning process pertaining to power reliability and resilience.

Decisionmakers at this campus determine the relative value of investing in a project by the economic effects if an event were to occur in the absence of the investment. To illustrate this concept, we will use a notional example. If, for example, $25,000 worth of food would be lost in an extended power disruption, then the decisionmakers would compare the value of $25,000 of spoiled food with the cost of installing and operating backup generation sources for the cafeterias. We had expected that these priorities might have to be compared with the priorities of the campus's hospital, but the hospital conducts its own energy planning, separate from the rest of campus. However, these power needs would be compared against the needs of other administrative functions, classes, campus housing, and research facilities. We learned that this largely a subjective process, and that priorities are set based on value judgements.

Besides upgrades to the campuswide microgrid, investments in power resilience are predominantly determined for spaces within buildings. For example, a specific research laboratory may require uninterrupted access to power to meet an experimental requirement. The head researcher of this laboratory would work with the campus energy group and determine whether the cost (shouldered by the lab's resources) is worth the risk to the experiments. These investments are rarely coordinated across labs or departments. Additionally, the campus does not have a standardized process for or method for valuing these energy assurance investments.

The university system does have a mechanism for considering investments at the systemwide level. Each campus self-selects projects, which are racked and stacked at the systemwide level, but it does not appear that there is any standard energy valuation process used to prioritize these projects. We learned that very few projects are funded through this mechanism; most energy investment decisions are made at the campus level. This process somewhat mirrors the Air Force AFCAMP process, although the Air Force prioritization model does attempt to prioritize projects based on a set of standard criteria.

Electric Utility Sector

Electric utilities' mission is to provide reliable power to consumers. Utilities generally serve a region and are subject to regulations set by public utility commissions (PUCs). Infrastructure investments and initiatives related to reliability and resilience are required to be brought to the PUC in a process called a *rate case*. Rate cases are the means by which a utility communicates its needs to the PUC and provides an analytical justification for its request. Rate cases justify increasing the rates paid by customers for investments, such as expanding generating capacity or increasing reliability in its transmission or distribution system.

Lawrence Berkeley National Laboratory (LBNL) researchers examined proposed investments in resilience and reliability by utilities through interviews with PUCs in Washington, D.C., California, and Florida and reported the following findings:

- Investments for reliability/resilience by utilities were predominantly included as cost recovery in the general rate case (rate case) proceedings.
- Little distinction was presented between reliability and resilience.
- The costs of investments related to reliability and resilience can be monetized, but the benefits are difficult to monetize.
- Political momentum, in addition to economic analyses, can propel reliability and resilience investments (LaCommare, Larsen, and Eto, 2017).

To build on these findings, we reviewed their evidentiary support for Washington, D.C., California, and Florida, and conducted an additional review of rate proceedings across other states to determine whether these findings were broadly applicable and whether other insights could be gathered from the full collection of rate cases.

Generally, we find that utilities communicate the value of energy assurance through general rate cases to their PSCs, consistent with the LBNL research findings. However, the electricity sector has yet to identify metrics for resilience and instead falls back on its standard metrics of reliability. For example, Commonwealth Edison asked the Illinois PUC for approval to place a line underground as its least cost option for 4.3-mile, 345 kV transmission line in Chicago for the purpose of enhancing reliability.[5] Additionally, the level of reliability is often assessed only with respect to past events and not a broader set of possible future scenarios.

The maturity and complexity of methodologies range across state PUCs. Few PUCs have developed robust forecasting models. States with legislation and executive branch initiatives on resilience appear to have more rate cases involving requests for cost recovery of resilience investments. For example, in a case brought to the New York State PUC, Consolidated Edison proposed a three-year initiative, including a model to assess the costs and benefits of resilience initiatives by comparing the monetary value associated with economic loss reduction pre- and post-investment (in units of $/kWh) (New York Department of Public Service, 2014). These findings were supported through an interview we conducted with a utility. The utility indicated that the rate increases to support reliability and resilience investments are dependent on the utility's ability to show cost-effectiveness.

---

[5] Commonwealth Edison used a computer model that included expected system reinforcements, new customer interconnections, generation supply, and forecasted loads (which include effects of energy efficiency programs) to run through different situations where certain elements become unavailable; the model results suggested the grid needed reliability and resilience reinforcements.

Catastrophic Risk Insurance

Much of the focus of disaster risk insurance literature pertains to constraints on the supply side that disrupt the functioning of the market. For example, Jaffee and Russell (1997) describe how a variety of institutional features, such as accounting practices, regulatory constraints, and tax provisions, make it difficult for the insurance market to create products that insure against catastrophic risk. In the catastrophe risk market, in the absence of concrete information on net economic and social benefits, many policymakers are reluctant to commit significant funds for risk reduction. However, when a disaster occurs, they are pressured into providing funds to assist victims and aid the recovery process (Michel-Kerjan and Volkman-Wise, 2011).[6]

Insurance Premiums Should Approach Expected Losses

A basic principle of all insurance markets is that insurance premiums should be roughly equal to expected losses. If premiums were well below expected losses, insurance companies would typically face revenue shortages in paying claims, so they would not provide coverage. From the demand side, if premiums are considerably greater than expected losses, individuals would typically do better by not paying for insurance and taking those losses if they happen.[7]

Expected loss is defined as the sum of the costs of a loss times the probability of that loss occurring, over different types of losses. One challenge in applying this expected loss principle to disaster risk insurance markets is that this expectation can be difficult to quantify. For example, in a national security context, if losses from a terrorism event are truly catastrophic, then the expectation approaches infinity. This would mean that any insurance premium, no matter how large, would be worth the cost.

A separate complicating factor is that the probability distribution for "bad events" is difficult to know. Despite considerable progress in recent years in identifying and measuring flood risks, for example, people make decisions about purchasing disaster insurance based on their subjective assessment of that probability distribution (Michel-Kerjan, 2010). As discussed in Appendix A, because individuals have biases in perceiving risks, there is no reason to expect that people's beliefs about the probability distribution for adverse events are reasonable or accurate. This can create wedges between what people may be willing to pay for insurance and the price they can obtain in an insurance market.

---

[6]   Another example of this comes from the international development space, where bilateral and multilateral donors allocate 98 percent of their disaster management funds to relief and reconstruction, but only 2 percent to proactive disaster risk management (Mechler, 2005).

[7]   Risk preferences can modify the extent to which insurance premiums can differ from expected losses. If an individual is completely risk-neutral, then from the demand side, premiums can be no higher than expected losses. The more risk-averse an individual is, the more he or she will be willing to pay premiums that exceed expected losses.

Despite these challenges, the expected loss principle could be an ideal starting point for thinking about valuation in disaster insurance markets (Dixon, 2014) and generally in commercial markets. But the principle falls short when it comes to planning for low-probability, high-consequence events such as catastrophic power outages, in the national security context or otherwise.

Exceedance Probability Curve

Hochrainer-Stigler et al. (2011) describe a series of metrics that might be useful in certain cases for thinking about valuing investments to reduce disaster risk. One concept is the *exceedance probability curve*, which is a relationship that indicates the probability p that at least X dollars is lost in a given year. In Figure B.1, we draw hypothetical exceedance probability curves for an Air Force installation. In any given year, the blue line, which depicts a baseline scenario, indicates that there is a 5 percent chance of at least a $500 loss. Losses cannot exceed $50,000, the total value of the installation, so the probability of a loss being greater than $50,000 is zero. In between, the curve slopes downward, as increased losses of at least X become less likely. Note that the area under the curve is the average annual loss, or the expected loss.

Investments in power resilience could shift this exceedance probability curve to the left. An example of this shift is depicted in the red line of Figure B.1, which is the new exceedance probability curve after the Air Force installation has made a power resilience investment. To evaluate whether different types of power resilience investments are worthwhile, one could simply compare the changes in expected losses that result from such investments with the expected annual costs of those investments.

Figure B.1
Example of the Exceedance Probability Curve

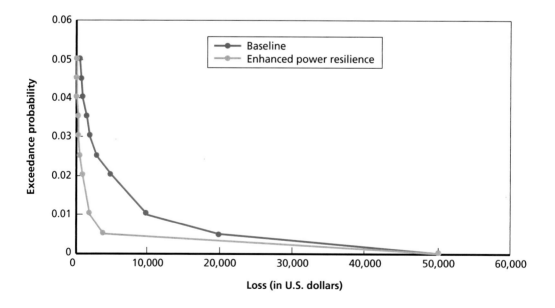

Hochrainer-Stigler et al. (2011) use exceedance probability curves to evaluate different measures to reduce the hurricane risk for homes in St. Lucia, flood risks for homes in Jakarta, earthquake risks for settlements in Istanbul, and flood risk for homes in Uttar Pradesh, India. Note that one key difference between applying this approach to valuing investments that protect against disaster risks for homes is that housing prices are readily available and relatively easy to measure.

In summary, none of the above approaches and principles is wholly suitable for application to the Air Force, although the electric power sector's treatment of resilience as an extension of reliability comes closest. The insurance industry's "avoided loss" principle to inform premium payments provides a partial solution but fails to incorporate the full array of costs associated with a failed or degraded Air Force mission and works poorly for the low-probability, high-consequence events. Similarly, exceedance probabilities are largely unknowable in the complex environment in which the Air Force is operating.

Upstream Processes for Setting and Modifying Requirements

In Chapter Five, we recommend that the Air Force incorporate power resilience into acquisition processes. For additional context, this appendix describes the JCIDS process, the DAS, and the Air Force specific processes that support the JCIDS and DAS where performance requirements[1] are developed. We focus on materiel solutions in this review for illustrative purposes but recognize that not all capability needs are materiel in nature and that these processes may vary based on the type of solution being evaluated. We then discuss the translation of performance requirements into power requirements at the end of the acquisition process and through the Air Force requirements modification process.

Development of Performance Requirements

Performance requirements for weapon systems are developed and validated through the JCIDS process and DAS (referred to here as the *acquisition process*). The JCIDS process supports the acquisition process by identifying, assessing, validating, and prioritizing capability[2] needs; and developing associated performance criteria (Chairman of the Joint Chiefs of Staff Instruction 3170.01I, 2015). The acquisition process is the mechanism by which DoD acquires weapon systems, information systems, and services to achieve goals described in strategic guidance documents, such as the National Security Strategy, the National Defense Strategy, the National Military Strategy, and the National Strategy for Homeland Defense; to meet the demands of various concept

---

[1] We use the term *performance requirement* in this appendix, consistent with Air Force terminology, which differs from *performance metric* and *targeted performance* introduced in Chapter Four. As discussed there, we do not suggest starting with a performance requirement, because the requirement may not always be known or it may not be the appropriate measure by which to assess performance using the proposed framework. This appendix is focused on the upstream processes that support requirements development.

[2] *Capability* or any derivation of the word in this appendix refers to "strategic capability," not an operational capability achievable in practice.

of operations (CONOPS) and mission planning documents; and support the U.S. Armed Forces (DoD Directive 5000.01, 2007; AcqNotes, 2015).[3]

Defense acquisition programs are classified by acquisition category (levels I–III) and type of program (e.g., Major Defense Acquisition Program [MDAP], Major Automated Information System [MAIS], and Major System) (DoD Instruction 5000.02, 2017). These categories, along with the Joint Staffing Designator[4] assigned to JCIDS documents, affect acquisition procedures, schedules, validation authorities, and the oversight required for the program.

Here, we describe the process of performance requirements development at the joint service level, recognizing that not all Air Force acquisition programs will be at this level. The Air Force requirements process supports the overarching joint guidance and mirrors the structure of the joint guidance. For programs that only require component level validation, the Air Force Requirements Oversight Council is the validation authority.

A simplified representation of the relationship of the JCIDS process and acquisition process is presented in Figure C.1.[5] In this figure, we include the key activities, outputs, and decision points[6] for a generic acquisition program where explicit or implicit valuation of electrical energy may occur or that may influence valuation or translation of requirements later in the process.

Starting at the top left of Figure C.1, the capabilities-based assessment (CBA) is the analysis portion of the JCIDS process. The CBA is meant to provide general recommendations on either a materiel solution, a non-materiel solution,[7] or a combination of both to identified capability requirements with capability gaps, in the context of a set of operational scenarios. Nonmateriel solutions are developed in parallel to materiel portfolios but separate from the acquisition process.

---

[3]   A third process, the PPBE process, rounds out the main components of the DoD decision support systems and is the process by which plans and programs are developed to determine resource requirements to satisfy the demands of strategic guidance documents. Because we are focused on performance requirements development in this section, we do not cover the PPBE process.

[4]   There are five Joint Staffing Designators. For details on these designators, see the *JCIDS Manual* (AcqNotes, 2015) and AFI 10-601 (2013).

[5]   This figure is meant to generally represent the JCIDS and acquisition processes. We note there may be variations on timing and sequence of certain steps in the process, the progression of requirements documents, and the stakeholders involved based on the nature of the capability requirements identified (e.g., a joint urgent operational need [JUON] versus deliberate acquisition), the type of gaps, and the proposed mitigation solutions. For more details on how these processes may vary, see the *JCIDS Manual* (AcqNotes, 2015).

[6]   The milestone A, B, and C decision points are included in the figure to orient the reader to the stage of the acquisition process.

[7]   Nonmateriel solutions fall into one of the eight elements of the DOTMLPF-P framework: doctrine, organization, training, materiel (e.g., equipment that does not require new development), leadership and education, personnel, facilities, and policy (AcqNotes, 2015).

Figure C.1
Key Activities, Outputs, and Decision Points in the JCIDS and Acquisition Processes

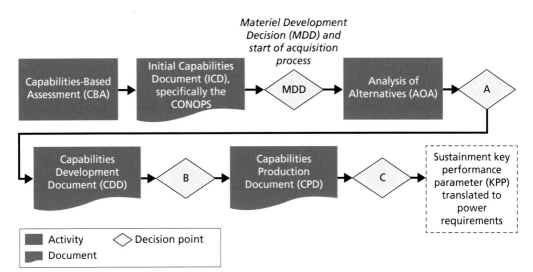

CBAs may be conducted by the services, the combatant commands (CCMDs), or other DoD components. When a gap is identified by a CBA, a sponsor will propose a new JCIDS document; in the Air Force, sponsors are typically a MAJCOM.[8] The intent of the CBA can be met through a combination of studies and analysis as deemed appropriate by the sponsor as long as requirements, gaps, and risks identified are traceable to an organization's role and mission, strategic guidance, and service or joint concepts (AcqNotes, 2015).[9] The required level of analytic rigor is dependent upon whether the capability is a previously validated solution or a new solution; the complexity of the mission; the uncertainties in future baselines, scenarios or CONOPS; and the consequences of operational failure (AcqNotes, 2015).

The CBA provides the analytic framework for establishing key performance parameters (KPPs) and key system attributes (KSAs), which becomes important later in the process for translating performance requirements into power requirements. We cover this further when we discuss the sustainment KPP and the process by which performance requirements are translated to power requirements in the next section.

---

[8]   When capabilities gaps are identified by a CCMD, the normal acquisition process requires a Service-level component to be the sponsor. If the requirement is filled through the Quick Reaction Capability Process (only an option for urgent operational needs), then a service-level component sponsor is not required (DoD Instruction 5000.02, 2017).

[9]   Sources of Air Force requirements include capability-based planning activities associated with the development of core function master plans, CBAs, top-down direction from higher authorities to fill a warfighting need (e.g., CCMD needs), and science and technology (S&T) activities. See AFI 10-601, *Operational Capability Requirements Development* for more detail.

Following the process flow in Figure C.1, when an identified capability gap presents significant operational risk to the joint forces, the capability requirements can be submitted to the appropriate authority for review and validation in the JCIDS process via an ICD.[10] The ICD contains the operational view (i.e., the start of a CONOPS or Commander's concept) which describes how a mission will be executed and the capabilities that are required in order to execute that mission. The CONOPS really drives the performance requirements and, key to this research effort, the required mode of operation during an outage and the outage tolerance. The ICD must be general enough not to prescribe a particular solution but specific enough to enable a sufficient analysis of alternatives (AoA) at the start of the acquisition process. That is, the content of the ICD should be mission specific rather than system-specific. The CBA and the ICD are important for laying the foundation for developing the performance metric critical to Step 1a in the proposed valuation framework presented in Chapter Four.

A validated ICD is required for the Materiel Development Decision (MDD), the first major decision point in the JCIDS and acquisition processes. The MDD is the point at which it is decided that a materiel solution is needed to address a given capability gap. The MDD authorizes the DoD component to begin the acquisition process and move forward with analyzing alternatives. At this decision point, the director of Cost Assessment and Program Evaluation or the DoD component equivalent presents the AoA study guidance and the lead organization presents the study plan to the Milestone Decision Authority (MDA).[11] If the MDD is approved, then the MDA designates the lead component and determines the acquisition phase of entry. Typically, acquisition programs move into the Materiel Solutions Analysis (MSA) phase, where an AoA is conducted, but if the DoD component has already completed sufficient analysis such that a product or solution is already narrowly defined (and technology maturation supports) then it is possible to enter into a later phase in the acquisition process.[12] Each phase of the process has an associated decision point to authorize

---

[10] AcqNotes, 2015. ICDs are the most common starting point for documenting requirements, but an ICD may not be required in certain cases. For example, if a capability solution has already been demonstrated in an operational environment through a validated JUON, a joint emergent operational need, or an experiment, the sponsor may request an ICD and/or a CDD waiver and the Milestone Decision Authority (MDA), then determines the appropriate starting point in the process (AcqNotes, 2015).

[11] All major decisions in the acquisition process are made by the MDA. For MDAP and MAIS acquisitions, the Under Secretary of Defense for Acquisition, Technology, and Logistics acts as the MDA but may delegate the authority in accordance with DoD Instruction 5000.02. For non-MDAP and non-MAIS acquisitions, DoD components use analogous component specific processes. The Air Force determines the validation authority in accordance with AFI 10-601 (2013) and AFI 63-101/20-101 (2017).

[12] The acquisition process includes five phases. We have not depicted all of these phases in Figure A.1 because the figure is meant to only represent key activities, decision points, and outputs that are relevant to our energy valuation question. The five phases of the acquisition process are the MSA phase, Technology Maturation and Risk Reduction phase; the Engineering and Manufacturing Development (EMD) phase; the Production and Deployment phase; and post Milestone C, the Operations and Support (O&S) phase, which encompasses the full

entry into the next phase. These decision points are represented as yellow diamonds in Figure A.1 with the letters A, B, and C (referred to as Milestone A, B, and C) (DoD Instruction 5000.02, 2017).

During the MSA phase, identified capability gaps begin to get translated into system specific requirements in the form of KPPs and KSAs as the concept for the product that will be acquired is developed. However, KPPs and KSAs are not formalized until later in the process. A CONOPS/Operational Mode Summary (OMS)/Mission Profile (MS) is prepared, which drives the performance requirements and includes details on "operational tasks, events, durations, frequency, operating conditions and environment in which the recommended materiel solution is to perform each mission and each phase of a mission" (DoD Instruction 5000.02, 2017, p. 19). This phase of the acquisition process and the CONOPS/OMS/MS can inform Step 2 (Define Power Outage Scenarios) in the proposed valuation framework presented in Chapter Four. The MSA phase ends with the Milestone A review, where the MDA will decide on the next phase of the acquisition process based the analysis completed in the MSA phase and how mature the technology is for the product being acquired.

Following the process flow in Figure C.1, the CDD is generated between Milestone A and Milestone B. The CDD is important because this is where mission performance requirements are defined and communicated and where KPPs and KSAs are first formally documented. Some of the content of the CDD may look very similar to the ICD developed prior to entry into the acquisition process; for example the operational context, the threat summary and the capability discussion. However, the CDD includes sections on the acquisition program summary, development key performance attributes (KPPs and KSAs) and other important additional performance attributes (APAs), spectrum requirements, intelligence supportability, weapon safety assurance, technology readiness, nonmateriel (i.e., DOTMLPF-P) considerations, and program affordability (AcqNotes, 2015).

To support validation of the CDD, a systems engineering trade-off analysis is conducted which is used to refine KPPs and KSAs (DoD Instruction 5000.02, 2017). The activities conducted between Milestone A and Milestone B are meant to "reduce technology, engineering, integration, and life-cycle cost risk to the point that a decision to contract for EMD can be made with confidence in successful program execution for development, production, and sustainment" (DoD Instruction 5000.02, 2017, p. 21). Once the CDD has been validated, the program moves into Milestone B review, which is considered the official start of a program, is where investment resources are committed to the program.

---

operational capability (FOC) date. For more information on the purpose of and the full set of activities involved with each of the phases, see DoD Instruction 5000.02 (2017).

Following Milestone B, the CPD is generated. The CPD is the formal requirements document that guides all program activities.[13] The biggest difference between the CDD and CPD is refinement of the performance attributes previously identified and defined in the CDD. KPPs, KSAs, and APAs are communicated in a threshold/objective format; the *threshold* represents the minimum performance the system is expected to have to be operationally effective and the *objective value* is the achievable desired operational goal, but usually at a higher risk to cost and schedule (AcqNotes, 2015). The performance requirements established in the CDD and CPD should inform the performance metric used in the proposed valuation framework in Chapter Four. There is a set of mandatory KPPs for which all CPDs must address. Relevant to this research effort, we highlight the sustainment KPP and the energy KPP, which covers fuel and electric power.

The sustainment KPP comprises three key attributes: availability, reliability, and ownership cost. *Availability* refers to the material availability (i.e., the number of operationally available end items and total population) and operational availability (e.g., percentage uptime or aircraft availability). *Reliability* measures the expected failure rate for systems over a specified timeframe under certain conditions. *Ownership cost* is the total life cycle ownership cost and is used in making decisions regarding availability and reliability. While these attributes might not explicitly consider power, the performance thresholds and objective values set here have implications for the enabling infrastructure and power assets needed to meet those performance requirements. For example, while outage tolerance may not be explicitly stated, it can be inferred from availability and reliability minimum performance values and objective values.

The mandatory energy KPP was just recently introduced into the process, and, as such, was not applied in the requirements development process for many of the fielded Air Force systems. For this reason, we do not refer to the energy KPP in the translation process because it is a new requirement. We also discovered through interviews with Air Force personnel involved with translating performance requirements in CPDs to technical specifications, and ultimately power supply assets and infrastructure (e.g., program office engineers, CE personnel), that they were not aware of the energy KPP.[14] The energy KPP applies to all weapon systems for which operational reach may be impacted by "the balance of energy performance of the system and the provision of energy to the system," or where "protection of energy infrastructure or energy resources in the logistics supply chain" is required (AcqNotes, 2015). This is a step in the right direction to changing the culture in the Services where electricity is often an afterthought, or just assumed to be available. However, the formal guidance that describes the energy KPP appears to be focused primarily on fuel requirements of the platform

---

[13] The CPD replaced the operational requirements document.

[14] Communication with Air Force personnel at Peterson AFB, June 26, 2017, at Los Angeles AFB, July 25, 2017, and at Vandenberg AFB, July 26, 2017.

and fuel delivery logistics. As the Air Force continues to increase its reliance on electric power to execute missions, electric power demands should also be integrated into analysis supporting development of the energy KPP.

Once the CPD has been validated, the program moves into Milestone C review, where a decision is sought to move into production and deployment. Up to this point, analysis is performed by the sponsor or other organizations on the mission owner side of things, with little, if any, formal input from installation support organizations (e.g., CE). This has implications for efficient use of Air Force resources. It is common for CE personnel to be handed facility and power requirements from the mission owner that have not taken into consideration the larger base infrastructure or considered whether mission function interdependencies and redundancies exist at the base or at another location.[15]

Translating Performance Requirements into Power Requirements
Once performance requirements have been defined, these requirements are then translated into power requirements and, ultimately, assets and supporting infrastructure. For power, this means first determining the electrical load, uptime, reliability, etc., and second, how many backup generators, UPS systems, or other secondary sources of power are needed to meet those performance requirements. Translation of performance requirements into power requirements is directly influenced by planning assumptions made during development of the performance requirements or during the translation process (i.e., what scenarios were used to assess required capability performance).

For example, consider two missions with the same electrical load. Mission A must maintain normal operation through an outage, and mission B can tolerate some amount of downtime (as determined by the performance requirements). The capacity and potentially the type of asset or supporting infrastructure used to meet performance requirements for these two missions will be different: Mission A requires uninterrupted power sustained through an outage, whereas mission B does not. Further, the assumptions made during the translation process as to the duration and frequency of power disruptions will also influence the power requirements. If it is assumed outages typically last no more than three days, then backup power supply will need to be procured to support three days of FOC for mission A; for mission B, maybe one day, maybe two days, or maybe no backup power will be procured if performance requirements are such that the mission can be without power for three full days. However, what happens if the outage lasts longer than three days? Or, in the case that missions may have enough backup generation capacity, but generators rely on diesel fuel resupply, what happens if the supply of diesel fuel is interrupted?

Although we did not identify a formal process that governs the specific activity of translation, we know translation happens both in acquisition and in sustainment

---

[15] Communication with Air Force personnel at the Space and Missile Center, Los Angeles AFB, July 25, 2017, and at Vandenberg AFB, July 26, 2017.

Figure C.2
Translating Performance Requirements in Acquisition

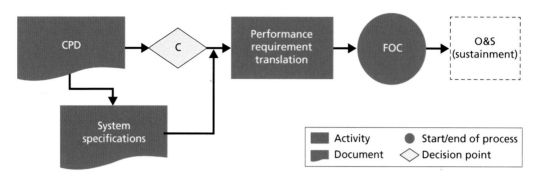

to meet performance requirements stated in the CPD (or other previously validated formal requirements document). We begin with translation during the acquisition process, which is conceptually represented in the process flow in Figure C.2. As with the previous figure, this diagram is meant to represent a generic acquisition program and only includes the key activities, inputs and outputs, and decision points relevant to this research effort.

At the top left, Figure C.2 starts with the CPD and Milestone C, which we described in the previous section. Coming out of the CPD and going into the performance requirement translation box, post Milestone C, is the system specifications document. The systems specifications document is derived from the CPD by engineers in the program office and is needed as an input into the performance requirement translation step because it takes the requirements expressed in terms of capabilities and translates them into requirements expressed in terms of technical systems specifications.[16] Then, engineers in the program office, in collaboration with a number of stakeholders, including MAJCOM engineers, mission partners, contractors, and installation CE personnel, use the system specifications to derive power requirements. They do this using a combination of information and methods, including the system specifications document, site suitability analysis, engineering assessments, and sometimes some discussion of frequency and duration of outages at that installation.[17] There is a formal requirement outlined in Air Force Policy Document 90-17 (2016) that the Air Force "be able to power any infrastructure identified as critical to the performance of mission essential functions independent of the utility grid *for the period of time needed to relocate the mission or for at least seven days, whichever is longer.*" However, it is not clear in some situations exactly who is responsible to plan for and fund that requirement.

---

[16] Communication with Air Force personnel at the Space and Missile Center, Los Angeles AFB, July 25, 2017.

[17] Interviews with subject-matter experts suggest there is no formal treatment of installation specific outage data at this stage in the process.

In conversations with subject-matter experts, we discovered that there is no set way to translate performance requirements into power requirements; it depends on the mission, the stakeholders, site specific architecture, and mission risk tolerances. We also discovered that power requirements may be expressed in several ways. For example, at Vandenberg AFB, while backup power requirements for range safety equipment requirements are expressed in terms of time, backup power requirements at one of the space launch complexes are expressed in terms of assets.

Once power requirements have been determined, assets have been procured, and the acquisition program meets the criteria for FOC, this ends the acquisition process and the program moves into the O&S phase, or sustainment phase.

While we did not identify a formal process that governs translation, the Air Force does have a formal process to document modifications to capability requirements for fielded systems that do not meet the minimum criteria for the JCIDS process (see *JCIDS Manual* [AcqNotes, 2015] for these criteria). These modifications can be documented using the Air Force Form 1067, *Modification Proposal* (AFI 10-601, 2013). A simplified version of this process is shown in Figure C.3.

The Air Force defines *modifications* as "changes to hardware or software to satisfy an operational mission requirement by removing or adding a capability or function, enhancing technical performance or suitability, or changing the form, fit, function, and interface of an in-service, configuration-managed AF asset" (AFI 63-101/20-101, 2017). Temporary or permanent sustainment modifications are typically initiated by a

Figure C.3
Translating Performance Requirements in Sustainment

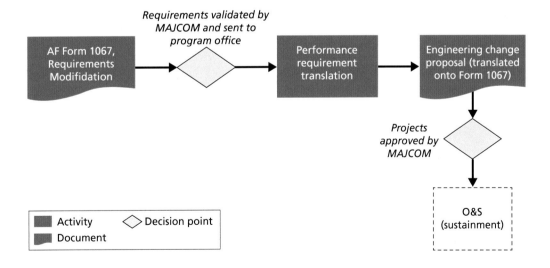

mission owner using Air Force Form 1067. The form is sent to the parent MAJCOM[18] to validate the requirement (and sometimes begin to identify funding sources) before sending it to the program office to initiate a technical evaluation and begin determining sustainment support needs and developing cost estimates.

The program manager works with the same stakeholders we discussed in the translation process in acquisition (e.g., program office engineers, installation CE personnel, MAJCOM engineers, mission partners, and contractors) to determine suitable options to fulfill requirements modifications. This process involves the same kind of analysis performed in the acquisition process—for example, site suitability analysis, engineering assessments, life cycle cost assessments. When modifications require larger infrastructure projects to support modified requirements, the IC and the Facilities Board must approve any changes to base asset management plans and activity management plans.

Finally, once a solution has been identified, typically an engineering change proposal is submitted to the program office, which then gets translated on to Air Force Form 1067. The MAJCOM is the final review and approval authority.

In principle, the valuation framework proposed in this report could be applied to both the acquisition and sustainment processes. Application of a test case would be most helpful in identifying any modifications that would be needed in these contexts.

---

[18]   In host-tenant situations, sometimes more than one MAJCOM will need to validate the form.

References

618th Air Operations Center, "About Us," webpage, undated. As of December 31, 2018: http://www.618tacc.amc.af.mil/About-Us/

AcqNotes, *Manual for the Operation of the Joint Capabilities Integration and Development System (JCIDS)*, Washington, D.C., February 12, 2015, including errata as of December 18, 2015. As of October 29, 2018: http://www.acqnotes.com/wp-content/uploads/2014/09/Manual-for-the-Operationsof-the-Joint-Capabilities-Integration-and-Development-System-JCIDS-18-Dec-2015.pdf

AFI—*See* U.S. Air Force Instruction.

Air Force Civil Engineer Center Planning and Integration Directorate, *FY19–21 AFCAMP Playbook*, 2017.

Air Force Civil Engineer Center Planning and Integration Directorate, *FY17–18 AFCAMP Playbook*, Washington, D.C., August 14, 2015. As of October 26, 2018: https://buildersummit.com/uploads/FY17-18_AFCAMP_Playbook_Consolidated.pdf

Air Force Instruction 10-601, *Operational Capability Requirements Development*, Washington, D.C., November 6, 2013.

Air Force Instruction 32-1021, Incorporating Change 1, *Planning and Programming Military Construction (MILCON) Projects*, Washington, D.C., December 29, 2016.

Air Force Instruction 32-1023_AFGM2017-01, *Designing and Constructing Military Construction Projects*, Washington, D.C., July 20, 2017.

Air Force Instruction 32-1032, *Planning and Programming Appropriated Fund Maintenance, Repair, and Construction Projects*, Washington, D.C., September 24, 2015.

Air Force Instruction 63-101/20-101, *Integrated Life Cycle Management*, Washington, D.C., May 9, 2017.

Air Force Instruction 65-501, *Economic Analysis*, Washington, D.C.: Department of the Air Force, October 29, 2018.

Air Force Office of Energy Assurance, "Frequently Asked Questions," June 2018. As of February 20, 2019: https://www.safie.hq.af.mil/Portals/78/documents/FAQs_Website_06-28-2018_FINAL.pdf?ver=2018-06-28-122239-383

Air Force Policy Directive 90-17, *Energy and Water Management*, Washington, D.C., November 18, 2016.

Barberis, N., "The Psychology of Tail Events: Progress and Challenges," *American Economic Review Papers and Proceedings*, Vol. 103, 2013, pp. 611–616.

Barseghyan, L, F. Molinari, T. O'Donoghue, and J. Teitelbaum, "The Nature of Risk Preferences: Evidence from Insurance Choices," *American Economic Review*, Vol. 103, No. 6, 2013, pp. 2499–2529.

Bayot, J., "Payouts Hinge on the Cause of Damage," *New York Times*, August 31, 2005.

Bedell, Darren, and Jim Norman, "NASA Launch Services Overview to NASA Advisory Council," National Aeronautics and Space Administration, slideshow, July 27, 2015. As of October 29, 2018: https://www.nasa.gov/sites/default/files/files/6-Norman-Launch-Services-Overview-Norman-Bedell-NAC-July-27-2015-Launch-Services.pdf

Boyle, K. J., "Introduction to Revealed Preference Methods," in P. A. Champ, K. J. Boyle, and T. C. Brown, eds., *A Primer on Nonmarket Valuation: The Economics of Non-Market Goods and Resources*, Vol. 3, New York: Springer, 2003.

Brown T. C., "Introduction to Stated Preference Methods," in P. A. Champ, K. J. Boyle, and T. C. Brown, eds., *A Primer on Nonmarket Valuation: The Economics of Non-Market Goods and Resources*, Vol. 3, New York: Springer, 2003.

Carron, Justin, *Emergency Power System Basics: Maintaining Always-On Power for Reliable Healthcare*, Cleveland, Ohio: Eaton, 2016. As of October 29, 2018: http://www.eaton.com/ecm/groups/public/@pub/@electrical/documents/content/wp083026en.pdf

Chairman of the Joint Chiefs of Staff Instruction 3170.01I, *Joint Capabilities Integration and Development System (JCIDS)*, Washington D.C., January 23, 2015.

Davis, Paul K., and Paul Dreyer, *RAND's Portfolio Analysis Tool (PAT): Theory, Method, and Reference Manual*, Santa Monica, Calif.: RAND Corporation, TR-756-OSD, 2009. As of December 5, 2018: https://www.rand.org/pubs/technical_reports/TR756.html

Davis, Paul K., Russell D. Shaver, and Justin Beck, *Portfolio-Analysis Methods for Assessing Capability Options*, Santa Monica, Calif.: RAND Corporation, MG-662-OSD, 2008. As of December 5, 2018: https://www.rand.org/pubs/monographs/MG662.html

Department of Defense Directive 5000.01, *The Defense Acquisition System*, Washington D.C, May 12, 2003, certified current as of November 20, 2007.

Department of Defense Instruction 5000.02, *Operation of the Defense Acquisition System*, Washington D.C., January 7, 2015, incorporating Change 3, August 10, 2017.

Dixon, Lloyd, *Catastrophic Risk in California: Are Homeowners and Communities Prepared?* Santa Monica, Calif.: RAND Corporation, CT-417, 2014. As of October 17, 2018: https://www.rand.org/pubs/testimonies/CT417.html

Drew, John G., Ronald G. McGarvey, and Peter Buryk, *Enabling Early Sustainment Decisions: Application to F-35 Depot-Level Maintenance*, Santa Monica, Calif.: RAND Corporation, RR-397-AF, 2013. As of October 17, 2018: https://www.rand.org/pubs/research_reports/RR397.html

Energy Information Administration, "Homes Show Greatest Seasonal Variation in Electricity Use," 2013. As of October 29, 2018: https://www.eia.gov/todayinenergy/detail.php?id=10211

Flannery, Jonathan, "Keeping the Power On: Survey Says EES Systems Highly Reliable." American Society for Healthcare Engineering, *Inside ASHE*, Winter 2014. As of October 29, 2018: http://www.ashe.org/compliance/ec_02_05_01/03/pdfs/EES-article-Inside-ASHE.pdf

Gerrity, Sarah, and Allison Lantero, "Infographic: Understanding the Grid," Washington, D.C.: U.S. Department of Energy, November 17, 2014. As of February 13, 2018: https://energy.gov/articles/infographic-understanding-grid

Gerstein, Daniel M., James G. Kallimani, Lauren A. Mayer, Lelia Meshkat, Jan Osburg, Paul K. Davis, Blake Cignarella, and Clifford A. Grammich, *Developing a Risk Assessment Methodology for the National Aeronautics and Space Administration*, Santa Monica, Calif.: RAND Corporation, RR-1537-NASA, 2016. As of October 25, 2017:
https://www.rand.org/pubs/research_reports/RR1537.html

Gigerenzer, G., "Out of the Frying Pan into the Fire: Behavioral Reactions to Terrorist Attacks," *Risk Analysis*, Vol. 26, No. 2, 2006, pp. 347–51.

Gilovich, T., D. Griffin, and D. Kahneman, eds., *Heuristics and Biases: The Psychology of Intuitive Judgment*, New York: Cambridge University Press, 2002.

Goodnough, A., "As Hurricane Season Looms, States Aim to Scare," *New York Times*, May 31, 2006.

Hochrainer-Stigler, Stefan, Howard Kunreuther, Joanne Linnerooth-Bayer, Reinhard Mechler, Erwann Michel-Kerjan, Robert Muir-Wood, Nicola Ranger, Pantea Vaziri, and Michael Young, *The Costs and Benefits of Reducing Risk from Natural Hazards to Residential Structures in Developing Countries*, Philadelphia, Pa.: University of Pennsylvania, 2011. As of December 7, 2018:
http://opim.wharton.upenn.edu/risk/library/WP2011-01_IIASA,RMS,Wharton_DevelopingCountries.pdf

Jaffee, D., and T. Russell, "Catastrophe Insurance, Capital Markets, and Uninsurable Risks," *Journal of Risk and Insurance*, Vol. 64, No. 2, 1997, pp. 205–230.

Judson, N., A. L. Pina, E. V. Dydek, S. B. Van Broekhoven, and A. S. Castillo, *Application of a Resilience Framework to Military Installations: A Methodology for Energy Resilience Business Case Decisions*, Lexington, Mass.: Lincoln Laboratory, Massachusetts Institute of Technology, October 4, 2016. As of October 26, 2018:
http://www.dtic.mil/dtic/tr/fulltext/u2/1024805.pdf

Kahneman, D., *Thinking, Fast and Slow*, New York: Farrar, Straus and Giroux, 2011.

Kress, Marin M., Katherine F. Touzinsky, Emily A. Vuxton, Bari Greenfeld, Linda S. Lillycrop, and Julie D. Rosati, "Alignment of U.S. ACE Civil Works Missions to Restore Habitat and Increase Environmental Resiliency," *Coastal Management*, Vol. 44, No. 3, 2016, pp. 193–208.

Kunreuther, H., R. Ginsberg, L. Miller, P. Sagi, P. Slovic, B. Borkan, and N. Katz, *Disaster Insurance Protection: Public Policy Lessons*, New York: John Wiley, 1978.

LaCommare, Kristina, Peter Larsen, and Joseph Eto, *Evaluating Proposed Investments in Power System Reliability and Resilience: Preliminary Results from Interviews with Public Utility Commission Staff*, Berkeley, Calif.: Energy Analysis and Environmental Impacts Division, Lawrence Berkeley National Laboratory, January 2017.

LaTourrette, Tom, and Henry H. Willis, *Using Probabilistic Terrorism Risk Modeling For Regulatory Benefit-Cost Analysis: Application to the Western Hemisphere Travel Initiative Implemented in the Land Environment*, Santa Monica, Calif.: RAND Corporation, WR-487-IEC, 2007. As of October 18, 2018:
https://www.rand.org/pubs/working_papers/WR487.html

Leftwich, James A., Robert S. Tripp, Amanda B. Geller, Patrick Mills, Tom LaTourrette, Charles Robert Roll, Jr., Cauley Von Hoffman, and David Johansen, *Supporting Expeditionary Aerospace Forces: An Operational Architecture for Combat Support Execution Planning and Control*, MR-1536-AF, Santa Monica, Calif.: RAND Corporation, 2002. As of February 18, 2019:
https://www.rand.org/pubs/monograph_reports/MR1536.html

Lichtenstein, S., P. Slovic, B. Fischhoff, M. Layman, and B. Combs, "Judged Frequency of Lethal Events," *Journal of Experimental Psychology and Human Learning*, Vol. 4, 1978, pp. 551–578.

Marquesee, Jeffrey, Craig Schultz, and Dorothy Robyn, *Power Begins at Home: Assured Energy for U.S. Military Bases*, Philadelphia, Pa.: Pew Charitable Trusts, Noblis, January 12, 2017.

Mechler, R., "Cost-Benefit Analysis of Natural Disaster Risk Management in Developing Countries," Eschborn: Deutsche Gesellschaft für Technische Zusammenarbeit (GTZ), 2005.

Mendoza-Hill, Alicia, "NASA Launch Services Program," National Aeronautics and Space Administration, slideshow, August 2015. As of October 29, 2018:
https://explorers.larc.nasa.gov/APSMEX/pdf_files/7_Launch_Vehicles_for_SMEXs-Aly_Mendoza.pdf

Michel-Kerjan, E., "Catastrophe Economics: The National Flood Insurance Program," *Journal of Economic Perspectives*, Vol. 24, No. 4, 2010, pp. 165–186.

Michel-Kerjan, E., and C. Kousky, "Come Rain or Shine: Evidence on Flood Insurance Purchases in Florida," *Journal of Risk and Insurance*, Vol. 7, No. 2, 2010, pp. 369–397.

Michel-Kerjan E., S. Lemoyne de Forges, and H. Kunreuther, "Policy Tenure Under the U.S. National Flood Insurance Program (NFIP)," *Risk Analysis*, Vol. 32, No. 4, 2012, pp. 644–658.

Michel-Kerjan, Erwann, and P. Slovic, eds., *The Irrational Economist: Making Decisions in a Dangerous World*, New York: Public Affairs, 2010.

Michel-Kerjan, E., and J. Volkman-Wise, "The Risk of Ever-Growing Disaster Relief Expectations," University of Pennsylvania Working Paper, 2011.

Mills, Patrick, Muharrem Mane, Kenneth Kuhn, Anu Narayanan, James D. Powers, Peter Buryk, Jeremy M. Eckhause, John G. Drew, and Kristin F. Lynch, *Articulating the Effects of Infrastructure Resourcing on Air Force Missions: Competing Approaches to Inform the Planning, Programming, Budgeting, and Execution System*, Santa Monica, Calif.: RAND Corporation, RR-1578-AF, 2017. As of February 18, 2019:
https://www.rand.org/pubs/research_reports/RR1578.html

Minkel, J. R., "The 2003 Northeast Blackout—Five Years Later," *Scientific American*, August 13, 2008.

Narayanan, Anu, Debra Knopman, James D. Powers, Bryan Boling, Benjamin M. Miller, Patrick Mills, Kristin Van Abel, Katherine Anania, Blake Cignarella, and Connor P. Jackson, *Air Force Installation Energy Assurance: An Assessment Framework*, Santa Monica, Calif.: RAND Corporation, RR-2066-AF, 2017. As of October 17, 2018:
https://www.rand.org/pubs/research_reports/RR2066.html

National Aeronautics and Space Administration, "NASA Online Directives Information System (NODIS) Library," webpage, undated. As of October 29, 2018:
https://nodis3.gsfc.nasa.gov/lib_docs.cfm?range=8

National Aeronautics and Space Administration, "Launch Services Program: Earth's Bridge to Space," Washington, D.C., 2012. As of October 29, 2018:
https://www.nasa.gov/sites/default/files/files/LSP_Brochure_508.pdf

National Aeronautics and Space Administration Policy Directive 8500.1C, *NASA Environmental Management*, Washington, D.C., December 2, 2013. As of October 29, 2018:
https://nodis3.gsfc.nasa.gov/displayDir.cfm?t=NPD&c=8500&s=1C

National Aeronautics and Space Administration Policy Directive 8610.24C, *Launch Services Program Pre-Launch Readiness Reviews*, Washington, D.C., 2005. As of October 29, 2018:
https://nodis3.gsfc.nasa.gov/displayDir.cfm?t=NPD&c=8610&s=24C

National Aeronautics and Space Administration Procedural Requirement 8570.1A, *NASA Energy Management Program*, Washington, D.C., July 12, 2013. As of October 29, 2018:
https://nodis3.gsfc.nasa.gov/displayDir.cfm?t=NPR&c=8570&s=1A

National Aeronautics and Space Administration Procedural Requirement 9420.1A, *Budget Formulation*, Washington, D.C., September 7, 2016. As of October 29, 2018:
https://nodis3.gsfc.nasa.gov/displayDir.cfm?t=NPR&c=9420&s=1A

New York State Department of Public Service, "Consolidated Edison (NY) Case No. 13-E-0030," last updated October 2014.

NGG Automatic Standby Generators, "Preventive Maintenance Agreement," webpage, undated. As of October 25, 2018:
http://nngenerator.com/preventive-maintenance-agreement/

North American Electric Reliability Corporation, *Reliability Standards for the Bulk Electric Systems of North America*, Atlanta, Ga., updated July 3, 2018. As of October 25, 2018:
https://www.nerc.com/pa/Stand/Reliability%20Standards%20Complete%20Set/RSCompleteSet.pdf

NPD—*See* National Aeronautics and Space Administration Policy Directive.

NPR—*See* National Aeronautics and Space Administration Procedural Requirement.

OEA—*See* Air Force Office of Energy Assurance.

Office of the Assistant Secretary of Defense, Energy, Installations, and Energy, *FY 2019/FY 2020 Energy Resilience and Conservation Investment Program and Plans for the Remainder of the Future Years Defense Program Guidance*, memorandum to the Assistant Secretary of Defense (Health Affairs), Assistant Secretary of the Army (Installations, Energy, and Environment), Assistant Secretary of the Navy (Energy, Installations, and Environment), Assistant Secretary of the Air Force (Installations, Environment, and Energy), and Directors of the Defense Agencies, Washington, D.C., September 15, 2017. As of October 29, 2018:
https://www.acq.osd.mil/eie/Downloads/IE/FY2019_FY2020%20ERCIP%20Guidance.PDF

Palm, Risa I., Michael E. Hodgson, R. Denise Blanchard, and Donald I. Lyons, *Earthquake Insurance in California: Environmental Policy and Individual Decision-Making*, Boulder, Colo.: Westview Press, 1990.

Parra Electric, "Generator Maintenance Service Agreement," webpage, undated. As of October 25, 2018:
http://parraelectric.com/wp-content/uploads/2015/06/Generator-Svc-Agreement_20140418.pdf

Public Law 115-91, National Defense Authorization Act of Fiscal Year 2018, December 12, 2017.

Samaras, Constantine, Rachel Costello, Paul DeLuca, Stephen J. Guerra, Kenneth Kuhn, Anu Narayanan, Michael Nixon, Stacie L. Pettyjohn, Nolan Sweeney, Joseph Vesely, and Lane F. Burgette, *Improvements to Air Force Strategic Basing Decisions*, Santa Monica, Calif.: RAND Corporation, RR-1297-AF, 2016. As of December 31, 2018:
https://www.rand.org/pubs/research_reports/RR1297.html

Sandroni, A., and F. Squintani, "Overconfidence, Insurance, and Paternalism." *American Economic Review*, Vol. 97, No. 5, 2007, pp. 1994–2004.

Schogol, Jeff, and Oriana Palwyk, "Incirlik Has Power Again, but Turkey Mission Faces Uncertain Future," *Military Times*, July 22, 2016.

Slovic, P., *The Perception of Risk*, London: Earthscan, 2000.

Sunstein, Cass R., "Probability Neglect: Emotions, Worst Cases, and Law," *Yale Law Journal*, Vol. 112, No. 61, 2002, pp. 61–107. As of December 7, 2018:
https://chicagounbound.uchicago.edu/cgi/viewcontent.cgi?article=12421&context=journal_articles

Taleb, N., *The Black Swan: The Impact of the Highly Improbable*, New York: Random House, 2007.

Taleb, N., D. Goldstein, and M. Spitznagel, "The Six Mistakes Executives Make in Risk Man," *Harvard Business Review*, October 2009.

U.S. Department of Energy, Office of Electricity Delivery and Energy Reliability, *Comparing the Impacts of Northeast Hurricanes on Energy, Infrastructure*, Washington, D.C., April 2013.

U.S. Department of Health and Human Services, Office of the Inspector General, *Hospital Emergency Preparedness and Response During Superstorm Sandy*, Washington, D.C., September 2014. As of October 29, 2018:
https://oig.hhs.gov/oei/reports/oei-06-13-00260.pdf

Zetter, Kim, "Inside the Cunning, Unprecedented Hack of Ukraine's Power Grid," *Wired*, March 3, 2016.